A Scientific Strategy for U.S. Participation in the

GOALS

(Global Ocean-Atmosphere-Land System)

Component of the CLIVAR

(Climate Variability and Predictability) Programme

Global Ocean–Atmosphere–Land System Panel
Climate Research Committee
Board on Atmospheric Sciences and Climate
Commission on Geosciences, Environment, and Resources

National Research Council

NATIONAL ACADEMY PRESS
Washington, D.C. 1998

This material is based upon work supported by the National Oceanic and Atmospheric Administration under Contract No. 50-DKNA-5-00015. Any opinions, findings, and conclusions or recommendations expressed in this publication are those of the author(s) and do not necessarily reflect the views of the above-mentioned agency.

Additional copies of this report are available from:
National Academy Press
2101 Constitution Avenue, N.W.
Box 285
Washington, D.C. 20055
800-624-6242
202-334-3313 (in the Washington, D.C., metropolitan area)

International Standard Book Number 0-309-06145-8

Cover: Long-term average August precipitation compiled by John M. Wallace, Todd P. Mitchell, and Alexis K.-H. Lau at the University of Washington's Joint Institute for the Study of the Atmosphere and Ocean (JISAO). The land data are taken from the Legates and Willmott climatology (1990, Int. J. Climatology, 10, 111-127), which is based on the historical record of rain gauge measurements. The ocean precipitation estimates are from the Microwave Sounding Unit (MSU) (Spencer, 1993, J. Climate, 6, 1301-1326). The climatology is based on averages for the period 1979 to 1992. The data used for the cover are also incorporated in Figure 3-2. The data were obtained from the following web site: *//tao.atmos.washington.edu/legates_msu/index.html* *iii*

v

Preface

The 10-year Tropical Ocean and Global Atmosphere (TOGA) program was a major and successful element of the World Climate Research Programme (WCRP), with participation by the United States and many other countries. TOGA demonstrated, for the first time, that with improved observations of the tropical Pacific Ocean and models that included interactions between the atmosphere and the ocean, aspects of El Niño/Southern Oscillation (ENSO) were predictable as much as a year or more in advance. This understanding, which was achieved around the mid-point of TOGA, led to the international TOGA field experiment "Coupled Ocean–Atmosphere Response Experiment" (TOGA-COARE) during which detailed measurements of atmospheric and oceanic properties were obtained. They are still being analyzed and used to improve the parameterization of processes in models. In the United States, the TOGA program also provided the initial impetus to establish a program on "Seasonal-to-Interannual Prediction" using the data provided by the TOGA observing system.

A second important outcome of TOGA was the realization that the scientific issues involved in seasonal-to-interannual prediction were more complex than originally foreseen. Climate variations on seasonal-to-interannual time scales seemed to be linked to other tropical oceans and regions, as well as variations in extratropical sea surface temperatures and land surface properties.

Following a meeting between the TOGA Panel of the National Research Council (NRC) and the WCRP Scientific Steering Group of the international TOGA program in July 1990 (Kona, Hawaii), the TOGA Panel recommended that, to take full advantage of the scientific progress made in understanding the dynamics of the coupled atmosphere–ocean system, a follow-on program to

TOGA needed to be created to focus on global climate variability on seasonal-to-interannual time scales. Towards this end, the TOGA Panel organized a series of study sessions covering the Asian–Australian monsoons, air–sea interaction in the tropical Atlantic, and the role of extratropical sea surface temperature variations. Based on these study sessions, the TOGA Panel proposed to the NRC's Climate Research Committee (CRC) that a Global Ocean–Atmosphere–Land System (GOALS) program be initiated. GOALS was conceived as a program that would support the international 15-year CLIVAR (Climate Variability and Predictability Programme) program of the WCRP. The CRC formed a GOALS Steering Committee to explore the concept further and to involve a broader community of scientists. The GOALS Steering Committee planned the GOALS Study Conference (March 1993, Honolulu, Hawaii), attended by 110 scientists, to address a number of scientific questions. Based on the conclusions of the conference, the Steering Committee assisted the NRC in preparing the GOALS Science Plan (NRC, 1994a).

The GOALS Science Plan called for a 15-year (1995-2010) research program that builds on the success of TOGA. The plan proposed an expansion of observational, modeling, and process studies to include the possible influences of the global upper ocean and the time-varying land moisture, vegetation, snow, and sea-ice on seasonal-to-interannual climate variability and prediction. The plan also proposed an organizational structure for GOALS and its relationship with CLIVAR. It recommended a tripartite structure, with a project office (GOALS Project Office), a scientific oversight body (NRC, with its CRC and the GOALS Panel), and a group of participating federal agencies (NRC, 1994a). The federal agencies would be responsible for implementing GOALS through coordinated funding of research grants. An Interagency GOALS Project Office would serve as a focal point for the implementation of the national research effort, and the GOALS Panel (with oversight from the NRC's CRC) would provide scientific guidance and oversight for the program. The plan also anticipated that principal investigators and consortia of principal investigators would carry out much of the actual implementation of the GOALS scientific plans. As the needs of the program dictate, the GOALS Project Office would invite groups to prepare coordinated sets of research proposals designed to address specific objectives of GOALS. It was proposed that close coordination between GOALS and the international CLIVAR program be maintained through a formal link between the project offices for the two programs and an informal liaison between the GOALS Panel and the International CLIVAR Scientific Steering Group.

In order to complete a retrospective assessment of TOGA, the CRC requested the TOGA Panel to summarize the findings of TOGA, its accomplishments, and its shortcomings with an emphasis on U.S. contributions (NRC, 1996). The assessment involved a broad community of scientists. This initial planning for U.S. participation in the new GOALS program, an ambitious attempt to extend our knowledge of the El Niño/Southern Oscillation and other short-term variations of climate, incorporates the conclusions of this assessment.

Following the development of the GOALS Science Plan, the GOALS Panel was tasked by the CRC to develop a strategy for U.S. participation in GOALS/CLIVAR. This report, which is the outcome of that effort, represents the culmination of extensive discussions and reviews between many members of the scientific community.

I wish to compliment the panel members in doing an outstanding job in raising and addressing numerous scientific, observational, logistical, and management issues. In addition to important sections on long-term observations, process studies, empirical and diagnostic studies, modeling, and data management, a section concerning the interaction between the physical science(s) communities and those involved in applications and human dimensions has been specifically incorporated, possibly for the first time in a science program. Thus the applications of climate forecasts and the manner in which climate impacts humans is considered to be an integral part of GOALS. The Executive Summary of this report encapsulates the key scientific issues regarding U.S. participation in GOALS/CLIVAR. We have no question in our minds that GOALS will significantly advance the state-of-science in seasonal-to-interannual climate prediction.

Peter J. Webster
Chairman, GOALS Panel

The National Academy of Sciences is a private, nonprofit, self-perpetuating society of distinguished scholars engaged in scientific and engineering research, dedicated to the furtherance of science and technology and to their use for the general welfare. Upon the authority of the charter granted to it by the Congress in 1863, the Academy has a mandate that requires it to advise the federal government on scientific and technical matters. Dr. Bruce M. Alberts is president of the National Academy of Sciences.

The National Academy of Engineering was established in 1964, under the charter of the National Academy of Sciences, as a parallel organization of outstanding engineers. It is autonomous in its administration and in the selection of its members, sharing with the National Academy of Sciences the responsibility for advising the federal government. The National Academy of Engineering also sponsors engineering programs aimed at meeting national needs, encourages education and research, and recognizes the superior achievements of engineers. Dr. William A. Wulf is president of the National Academy of Engineering.

The Institute of Medicine was established in 1970 by the National Academy of Sciences to secure the services of eminent members of appropriate professions in the examination of policy matters pertaining to the health of the public. The Institute acts under the responsibility given to the National Academy of Sciences by its congressional charter to be an adviser to the federal government and, upon its own initiative, to identify issues of medical care, research, and education. Dr. Kenneth I. Shine is president of the Institute of Medicine.

The National Research Council was organized by the National Academy of Sciences in 1916 to associate the broad community of science and technology with the Academy's purposes of furthering knowledge and advising the federal government. Functioning in accordance with general policies determined by the Academy, the Council has become the principal operating agency of both the National Academy of Sciences and the National Academy of Engineering in providing services to the government, the public, and the scientific and engineering communities. The Council is administered jointly by both Academies and the Institute of Medicine. Dr. Bruce M. Alberts and Dr. William A. Wulf are chairman and vice chairman, respectively, of the National Research Council.

Acknowledgment of Reviewers

This report has been reviewed by individuals chosen for their diverse perspectives and technical expertise, in accordance with procedures approved by the NRC's Report Review Committee. The purpose of this independent review is to provide candid and critical comments that will assist the authors and the NRC in making the published report as sound as possible and to ensure that the report meets institutional standards for objectivity, evidence, and responsiveness to the study charge. The content of the review comments and draft manuscript remain confidential to protect the integrity of the deliberative process. We wish to thank the following individuals for their participation in the review of this report:

Alan Clarke, Bedford Institute of Oceanography, Dartmouth, Canada
Chongyin Li, Chinese Academy of Sciences, Beijing, China
Neville Nichols, Bureau of Meteorology Research Centre, Melbourne, Australia
Soroosh Sorooshian, University of Arizona, Tucson, United States of America
Akimasa Sumi, University of Tokyo, Japan

While the individuals listed above have provided many constructive comments and suggestions, responsibility for the final content of this report rests solely with the authoring committee and the NRC.

Contents

Executive Summary

The National Research Council's (NRC's) study on the Global Ocean–Atmosphere–Land System (GOALS) began in 1995 with the establishment of the GOALS Panel, which was tasked by the Climate Research Committee (CRC) to recommend a strategy for U.S. participation in the international GOALS/Climate Variability and Predictability (CLIVAR) programme.

Specifically, under the oversight of the Board on Atmospheric Sciences and Climate (BASC), through its CRC, the panel was charged to:

• Encourage the initiation of a U.S. GOALS Project Office and provide overall scientific leadership and regular guidance on long-range scientific policy, planning, progress, and priorities to that office as well as to involved agencies on behalf of the CRC.

• Report regularly to inform the CRC on the Panel's involvement in U.S. GOALS plans and activities and to receive guidance from the CRC on GOALS matters in the context of the overall U.S. climate research program.

• Advise on U.S. participation and provide U.S. inputs to the international GOALS program planning.

The current document follows directly from the panel's charge. The principal purpose of the document is to provide a strategy for U.S. participation in GOALS/CLIVAR, which should endure for the 15-year duration of the program. At the same time, it is realized that during a program of such length, an evolution of emphases and priorities will occur with advances of knowledge and new discoveries. This document emphasizes explicit and necessary first steps and initial priorities.

The scientific objectives of GOALS are described in the GOALS Scientific Plan (NRC, 1994a) and summarized in Section 2 of this report. A key aspect of GOALS is the identification of those characteristics of the global system that would make significant improvements in climate prediction on seasonal-to-interannual time scales possible.

The impetus for GOALS stemmed from the success of TOGA, the Tropical Oceans and Global Atmosphere program, which established the existence of "predictability" on seasonal-to-interannual time scales in the Pacific Ocean for equatorial sea surface temperature (SST) and for precipitation around the tropical Pacific and in some regions far removed. Much of this predictability is attributed to ocean–atmosphere coupled processes in the Pacific Ocean basin. More exhaustive research is considered necessary to explore the possibility of recognizable interactive processes occurring within or between other components of the Earth system. For example, empirical studies have hinted that anomalous monsoon seasons are foreshadowed by coherent and robust signals in remote locations. Though not fully investigated, linkages are also apparent between the monsoons and El Niño/Southern Oscillation (ENSO), Eurasian snow cover, and other climate system components. If "climate memory" exists in the higher-latitude systems, it is hypothesized to be in the ocean, sea-ice, and hydrological and ground surface processes over the continents.

A fundamental objective of GOALS is to extend the domains of interest outwards from the tropical Pacific basin, the region of primary attention during TOGA. A compelling reason for this expanded domain is the need to understand and quantify the effect of fluctuations in other major heat sources and sinks of the tropics and subtropics on the global general circulation and thereby improve predictions of weather and climate in both local and remote regions—including those in higher latitudes. Thus, the GOALS Panel proposes extending the study of ocean–atmosphere interaction on seasonal-to-interannual time scales to the three tropical oceans. To this end, three initial regional observational foci have been identified in Section 3: (1) the tropical Pacific Ocean, (2) the Americas and surrounding oceans, and (3) the Indian Ocean and land masses that surround it. In addition to these regional foci and their remote influences on higher latitudes, the panel recommends that GOALS investigate predictability inherent to the extratropics, as well as the interaction between land-surface hydrological processes and the atmosphere at all latitudes.

The global extent of the program poses a considerable research and observational challenge. Section 6 details the observational component, which will require the full utilization of all available (and planned) resources in order to achieve global coverage of key climate variables. In this regard, it is emphasized that although significant and increasing problems exist in monitoring the atmosphere, far greater problems exist in monitoring the ocean. The necessity for continuous global observations of the ocean, land, ice, and atmosphere underscores the importance of a comprehensive satellite program for GOALS, in conjunction with in situ measurements. The key observational variables identified by the panel are also detailed in Section 6. Section 7 recommends a philosophy for the establishment of special process studies aimed at elucidating specific physical processes important for seasonal-to-interannual prediction. The observations to be made within the GOALS program also underpin empirical and diagnostic studies (Section 8) and provide the data for the initialization and evaluation of models. The program's modeling component, described in Section 9, is hierarchical and inherently global even though the development and use of "embedded" high-resolution models are also envisaged, particularly for the prediction of regional and local climate variability and its impacts on society, industry, and natural resources.

The GOALS Panel assigns very high priority to the maintenance and enhancement of the TOGA atmospheric and oceanic observing systems and continued investigation of ENSO, implemented under TOGA. Equal priority is assigned to the study of atmosphere–ocean–land interaction in the Americas and efforts to improve the understanding and prediction of the North and South American monsoon systems. This latter effort has been codified through the U.S. Pan American Climate Studies (PACS) program and with international partners as the Variability of the American Monsoon Systems (VAMOS) program. Next in order of priority is the study of the Asian–Australian monsoon system. The Asian–Australian monsoon system undergoes significant interannual variability, whose impact is important to a large proportion of the population of the planet and the global climate system. To ensure a meaningful scientific investigation of this component of GOALS, the panel recommends that the United States seek international partners for the exploration of the Asian–Australian monsoon system. Just as the maintenance of the Pacific ocean–atmosphere observational array and the PACS program are expected to require strong interagency support in the United States, the global and far reaching totality of GOALS would require a well-coordinated international effort. To this end, the GOALS Panel should work closely with the International CLIVAR Scientific Steering Group of the World Climate Research Programme (WCRP).

Furthermore, the GOALS Panel recognizes that achieving the scientific objectives of GOALS requires that significant attention be paid to events and phenomena of both shorter and longer time scales (i.e., time scales outside the seasonal-to-interannual range). Thus, ocean–atmosphere–land interactions on intraseasonal (weeks to months) time scales are considered important and worthy of serious investigation. Correspondingly, the interdecadal modulation of the interannual variability of the coupled system also needs to be clearly understood. To address these concerns the panel proposes that the scientific development and implementation of GOALS should be closely coordinated with the Global Energy and Water Cycle Experiment (GEWEX) and the Climate Variability on Decade-to-Century Time Scales (Dec-Cen) programs, thereby contributing to a cohesive and overarching structure consistent with the objectives of the U.S. Global Change Research Program (USGCRP) and international CLIVAR.

The prospect of predicting seasonal-to-interannual variations in different parts of the world brings with it considerable opportunities to benefit climate-sensitive social and economic activities such as those involved in the planning and management of agriculture, water, fisheries, energy, natural resources, public health, and tourism, among others. The panel believes strongly in the concept of an "integrated" prediction system that includes the physical and social sciences as active partners (Section 10). To this end, the panel places high priority on the establishment of an ongoing dialogue between the physical scientists who make climate observations and predictions, and the user community. This dialogue is likely to be most beneficial if commenced very early in the development of GOALS.

Data management is viewed as an overarching activity that cuts across all the other elements of GOALS (Section 11). Internal linkages described in Section 12 refer to possible implementation aspects comprising a combination of individual research investigators and consortia, the precise funding details of which should be determined by the federal agencies responsible for and involved in GOALS. The interactions necessary between GOALS and other national and international research and observational programs are also summarized in Section 12, as are other coordination mechanisms and implementation aspects.

To further develop more specific plans for the implementation of GOALS, the panel proposes the establishment of working groups on observations and modeling, in addition to the creation of a GOALS Project Office, and a coordinated interagency federal mechanism to which the panel could be a scientific advisory body.

1

Introduction

A broad range of empirical and modeling evidence suggests that there may be many predictable modes of fluctuation on seasonal-to-interannual time scales. One such mode, ENSO, a strong interannual variation in climate centered in the tropical Pacific, has been the subject of intensive research that has led to skillful and useful predictions. The NRC's publication on "Learning to Predict Climate Variations Associated with El Niño and the Southern Oscillation" elaborates, in considerable detail, current understanding on this subject (NRC, 1996). Once other predictable modes are identified and modeled, the societal benefit from improved forecasts utilizing information about these additional elements or modes will be increased.

At present, our most advanced understanding of climate system variations on seasonal-to-interannual time scales concerns the coupling between the ocean and atmosphere of the tropical Pacific Ocean. Consequently, it is coupled variations in this region that we are most successful at forecasting on such time scales. Over the past decade, the WCRP TOGA program helped produce significant advances in our understanding and ability to predict ENSO. Building on the success of TOGA, seasonal-to-interannual research and prediction efforts need to be expanded to cover the global domain and should include a concerted study of phenomena in addition to ENSO that impact short-term climate variability. This is the base objective of the GOALS program (NRC, 1994a). TOGA concentrated on the coupling between the tropical atmosphere and the ocean. Expanding on the scientific scope of TOGA, GOALS also considers the interactions between the land and the atmosphere and between the midlatitude ocean and the atmosphere, as well as interactions of snow cover over land, and oceanic ice fields with the atmosphere and the ocean. Land surface processes, including vegetation and the biosphere, are explicit elements of GOALS and are to be studied in conjunction with GEWEX.

U.S. GOALS is a contribution to the international CLIVAR programme (WCRP, 1995), comprising three major sub-programs: (1) GOALS, (2) DecCen, and (3) Anthropogenic Climate Change (ACC). To facilitate a planning process for research into climate variability on seasonal-to-interannual time scales, the NRC established the GOALS Panel. Under the oversight of BASC, through its CRC, the panel was charged to:

- Provide overall scientific leadership and regular guidance on long-range scientific policy, planning, progress, and priorities to the proposed U.S. GOALS Project Office and involved agencies on behalf of the CRC.
- Report regularly to inform the CRC on the panel's involvement in U.S. GOALS plans and activities and to receive guidance from the CRC on GOALS matters in the context of the overall U.S. climate research program.
- Advise on U.S. participation and providing the U.S. inputs to the international GOALS program planning.

The current document follows directly from the panel's charge. The principal purpose of the document is to provide a strategy for U.S. participation in GOALS/CLIVAR, which should endure for the 15-year duration of the program. At the same time, it is realized that during a program of such length, an evolution of emphases and priorities will occur as knowledge accumulates and new views develop. This document emphasizes explicit and necessary first steps and initial priorities.

As part of the planning process for U.S. GOALS, this report presents a strategy for pursuing the scientific objectives of the U.S. component of GOALS, as described in a previous report (NRC, 1995). Its purpose is to facilitate the development of implementation plans for GOALS while encouraging innovative strategies. It does this by setting forth the scientific objectives and basis of GOALS, clarifying the temporal and spatial focus of the program, and describing the extent and type of work envisioned under the six elements of U.S. GOALS described in Section 4.

2

Scientific Objectives and Basis of GOALS

The scientific questions, objectives, and justification for the GOALS program are discussed extensively in the U.S. GOALS Science Plan (NRC, 1995) and the CLIVAR Science Plan (WCRP, 1995). The scientific objectives of U.S. GOALS are to:

1. Understand global climate variability on seasonal-to-interannual time scales.
2. Determine the spatial and temporal extent to which this variability is predictable.
3. Develop the observational, theoretical, and computational means to predict seasonal-to-interannual variations.
4. Make experimental climate predictions on seasonal-to-interannual time scales.

These objectives build on the predictability that exists in the tropical Pacific Ocean associated with the coupled ocean–atmosphere processes of ENSO. Achieving them requires an enhanced understanding of seasonal-to-interannual variability in other regions. Improvements in exploiting predictability are expected as a result of developing new models that incorporate the knowledge gained from diagnostic and empirical studies in regions outside the tropical Pacific Ocean region. In particular, incremental additions to seasonal-to-interannual climate prediction skills are anticipated from the study of variations of other major tropical heat sources and sinks (e.g., the monsoon systems, the western Pacific warm pool, and the tropical American land masses) and determination of the impacts of these variations on other regions of the tropics, subtropics, and higher latitudes. In addition, the objectives call for assessment of the predictability that may be inherent in the coupled ocean–atmosphere–land–ice system of the extratropics and incorporation of these new elements into experimental prediction schemes. The ultimate aim of GOALS is to develop an operational global climate prediction capability on seasonal-to-interannual time scales. Progress towards this aim is expected to evolve on the basis of experimental prediction experience and by concerted efforts to produce better coupled ocean–atmosphere–land models.

The underlying physical principle that guides the modeling, empirical studies, process studies, and monitoring activities of GOALS is that water, in its various forms, provides thermal inertia and heat storage anomalies in the Earth's climate system and, through hydrological processes, defines the dominant interaction between components of the climate system. In turn, water's thermal inertia, varying heat transport, and varying heat storage provide "memory" and, therefore, predictability for the climate system. Thus, GOALS focuses on the coupled ocean–atmosphere–land–ice system, with particular emphasis on the hydrological cycle, oceanic heat storage and transport variations, and land surface processes as they pertain to the prediction of seasonal-to-interannual anomalies.

Of particular interest for the U.S. program is the assessment of predictability and development of skillful prediction methods for the North American sector, while encouraging similar activities in other regions where research shows there is potential predictive skill.

3

Temporal and Spatial Foci for GOALS

The principal temporal focus of GOALS is seasonal-to-interannual, and the breadth of its spatial interest is global. However, to accomplish its scientific objectives, the program also will have to consider temporal variability over a broad range from diurnal to decadal time scales. For the design and implementation of process studies, specific regions embedded in the global domain will require special emphasis.

The range of time scales of interest is shown in Figure 3-1. GOALS concentrates on phenomena that include the interannual (e.g., ENSO), biennial, and annual variability of the coupled ocean–atmosphere–land system, as well as intraseasonal fluctuations (e.g., monsoon break and active periods, and the Madden-Julian Oscillation) that act to create patterns of weather and may conceivably influence seasonal-to-interannual time scales. At the higher-frequency end of the spectrum and where land surface processes are more directly involved, GOALS plans on specific interfaces with GEWEX. Investigations on low-frequency (long-time period) phenomena within the GOALS program include the decadal modulation of and trends in ENSO, interannual variability of monsoon and rainfall patterns, and the seasonal and interannual variability of the North Atlantic Oscillation and the Pacific North American Oscillation. Of importance also are the climate signals and their long time scale modulations found in ice cores, tree rings, coastal corals, pollen, etc. For these studies, GOALS will require an interface with the DecCen program.

The GOALS program gives due recognition to the possibility that anthropogenic effects on climate may modulate seasonal-to-interannual variability of the coupled ocean–atmosphere–land–ice system. The consideration of the anthropo-

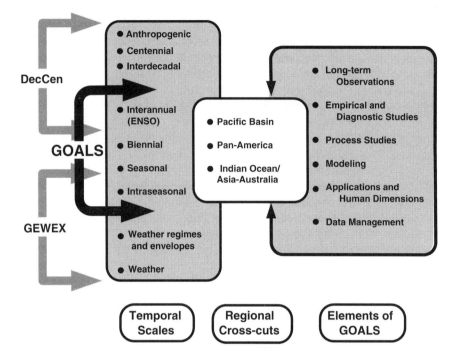

FIGURE 3-1 Temporal scales, regional foci, and elements of the GOALS program. Note that the overlapping time scales of interest between the GOALS program and those of GEWEX and DecCen implies the need for well-coordinated interfaces between the programs.

genic effects on climate and climate impacts is primarily handled by DecCen and ACC with which close coordination is planned. Combining the focus of GOALS (seasonal-to-interannual time scales) and those of GEWEX and DecCen, leads to a complete program that addresses key issues concerning the predictability of climate on all time scales—a primary concern of the USGCRP, and the international CLIVAR program of the WCRP.

The spatial focus of GOALS is, in effect, one of nested domains. The largest scale, global, is the domain for global models. At this scale, "global" observations are needed for the initialization of models and for quantifying and evaluating their prediction skills. Empirical and diagnostic studies are proposed to document the seasonal-to-interannual variability of the natural global system. Both modeling and data analysis investigations should search for evidence of relationships between sub-components of the Earth system and low-frequency variability that may be exploited for improving prediction capabilities. This last activity is sometimes referred to as prospecting for predictability and will have a bearing on which parts of the global domain are chosen for concentrated GOALS subprograms and projects.

Within the global domain, there are several regions in the tropics that will become foci of GOALS research. Within these regions process studies, long-term observations, and modeling efforts will target specific elements of predictability. The recommended strategy for U.S. GOALS for choosing the regional foci is defined below:

- The predictability found in the Pacific Ocean through ENSO should be followed to other regions of the tropics to determine how much of the variance can be explained in terms of ENSO.
- The regional foci should encompass the major tropical heat sources and sinks. There is considerable evidence of interannual variability in other regions of the tropics (e.g., the Asian–Australian monsoon system) together with predictive skill illustrated using empirical methods. Although there appears to be coupling between ENSO and variability in the other major heat sources and sinks, there also appears to be evidence of independent variability.
- The predictable elements found in the tropics from ENSO, or associated with the other heat sources and sinks in the tropics, should be projected to higher latitudes to see if they imply extratropical predictability. There is some evidence that the extratropics may not possess significant inherent predictability. However, this conclusion is based on limited observational studies and on coupled model experiments of relatively crude formulation, especially the oceanic component that has often been represented by fixed SST distributions. Investigations are needed to determine whether predictability on seasonal-to-interannual time scales, wherever it exists, has roots within the tropical system or whether there are predictable modes in the higher latitudes. To isolate predictable elements of higher latitudes (if they exist) it will be necessary to include more sophisticated land surface and sea ice processes than currently exist in climate models.

Figure 3-2 displays the regional foci now thought to be important for the attainment of the scientific objectives of GOALS. These are: (A) the tropical Pacific Ocean; (B) the Pan-American region extending from the eastern Pacific Ocean to Africa—the focus of the PACS program; and (C) the Asian–Australian monsoon system that encompasses the Indian Ocean, the western Pacific warm pool and the oceanic regions to its northwest and southwest, the oceanic throughflow through the Indonesian Archipelago that connects the Pacific and Indian Oceans, and the land masses of Australia, South Asia, and East Africa.

It is recommended that the initial emphasis for U.S. GOALS should be placed on the Pacific basin and the Pan–American region because of their significant and direct effects on North American climate. It is noted, however, that as our understanding of the global climate system develops, priorities may be expected to change and new regional foci identified.

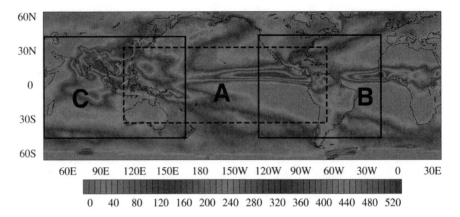

FIGURE 3-2   The three regional foci of GOALS: the tropical Pacific Ocean (A), the Americas (B), and the Indian Ocean and surrounding continental regions (C). Background shading represents the long-term August precipitation rates as estimated by the MSU (units: millimeters per month). Regional foci encompass major precipitation regions of the tropics.

The panel suggests that the extension of GOALS to the global domain and the expansion of present observational capabilities beyond the tropical Pacific to other regions should occur in an orderly manner and in parallel with a regionalization of prediction problems. Improvements in regional prediction skills are particularly important for applications and interfaces with society, both of which are high priorities of GOALS.

4

The Six Elements of GOALS

GOALS is a complex program calling for investigations of interaction between several major components of the climate system, possibly for the first time. Importantly, this interaction also implies the improvement of interfaces between the specialized scientific disciplines that previously concentrated on individual aspects of the global climate system. The program also demands a seamless interface between observing systems, models (and time integrated predictions), applications, and so forth. In recognition of the complexity of GOALS, the panel proposes the following six key elements or activities that need to be supported in order to achieve the objectives of the program:

1. long-term observations and analyses;
2. process studies;
3. empirical and diagnostic studies;
4. modeling;
5. applications and human dimensions; and
6. data management.

The first four elements, along with the last, were identified in the GOALS Science Plan (NRC, 1995). Because of the direct impact of seasonal-to-interannual climate prediction on society, applications and human dimensions have been added as an element of GOALS. The specific reason for this addition by the GOALS Panel is to support the concept of an end-to-end predictive capability that provides an interactive interface between the physical scientists who produce the forecasts and the user community.

The six elements of GOALS are highly interrelated, and each element has the potential to make an impact on the program's full geographical and phenomenological range. Thus, the program needs to be balanced with strong efforts in all six areas. Prioritization in each of these elements will be based on the objectives outlined at the end of Section 2.

Sustained long-term observations of key variables of the climate system form the backbone of GOALS. They are required in order to provide both a robust description of seasonal-to-interannual variability and its context relative to longer and shorter time scales. The data requirements for model initialization and verification often help in defining the observations and observing systems needed. Comprehensive analyses of these global data are required with the aim of producing global syntheses of climate and its variations in useful gridded format.

Empirical and diagnostic studies often rely on the long-term observations program element and analyzed data sets. They are invaluable in defining the areas in which model improvements are required. They also help assess which process studies are needed and help identify physical relationships which suggest aspects of the climate system that are potentially predictable.

Process studies are required to improve the predictive capability of coupled models by obtaining the data necessary for analyses, which can lead to explanations of the physical processes that need to be represented by new parameterization schemes. They need to have a sampling density sufficient to resolve time and space scales of variability in the region of interest. They are expected to lead to an explanation of the physical links that can extend predictable signals to remote locations. They are necessary to test hypotheses regarding the dynamics of interactive sub-components of complex systems. Process studies can also provide a physical linkage between the more widely spaced observations obtained in new long-term data sampling arrays and networks.

Modeling involves the development and application of improved complex coupled ocean–atmosphere and ocean–atmosphere–sea–ice–land–surface models. Modeling is considered the unifying theme underlying GOALS because of the ability of models to integrate or encapsulate knowledge derived from process studies, observations, and so forth, and to generate the products required to support the applications of relevance to societal problems. The results of studies with less complex models can suggest physical processes that should be incorporated into comprehensive coupled models; they should help the development of parameterization schemes for processes not explicitly resolved in complex models. Simplified models are also necessary to better understand the behavior of complex models. They can suggest regions and phenomena where predictability exists.

Applications and human dimensions studies enhance the utility of forecasts by taking into account user needs.

Data management is considered a critical element that cuts across the totality of GOALS. It should be designed to provide a fluid interaction with the scientific community at large. A carefully configured data management plan will provide a legacy for GOALS. The coordination of a GOALS data management plan with the international CLIVAR data management plan is essential.

Figure 4-1 provides a composite description of the GOALS program. The six elements are grouped into four components, to simplify their schematic repre-

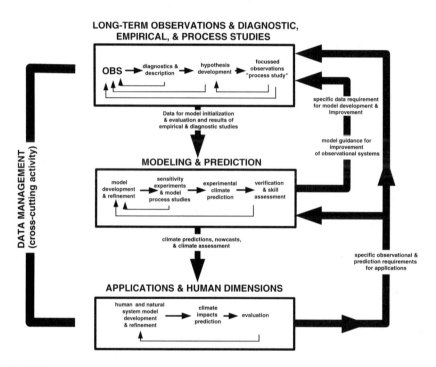

FIGURE 4-1 Composite description of the GOALS program. Three major components are described within the shaded boxes: (1) long-term observations and diagnostic, empirical, and process studies; (2) modeling and prediction; and (3) applications and human dimensions. Within each component, the arrows indicate a research sequence consistent with the scientific method, including feedbacks among various steps. Each of the components is connected through the provision of products (heavy arrows). However, there are feedbacks between the components (heavy arrows on right hand side). For example, models can be used to guide the improvement of observational networks. Models also will require specific observations to improve certain parameterizations and thus may promote observational process studies. Furthermore, the user community may require greater specificity in terms of model products or observations. Data management is seen as embracing all three components.

sentation. The first is associated with observation systems and their iterative development. The second refers to modeling and prediction, while the third concentrates on applications and human dimensions. The fourth, data management, is seen as an overarching element. Within each class of elements, a procedure for research and development is described. Depicted between the classes of elements are a series of interactions that signify the feedbacks considered necessary for the development of each component of GOALS.

5

Context for the U.S. GOALS Strategy

The U.S. GOALS strategy focuses on the improvement of seasonal-to-interannual climate prediction, especially over North America. Several other operational and research programs that have the capacity to assist in attaining this objective are also planned or are already in place. In developing the strategy for the U.S. participation in GOALS, the panel recognizes that it is important to coordinate with these other efforts and establish links with them. One of the most important institutions on which GOALS will rely is the operational system developed for weather monitoring and prediction. During TOGA, a skeletal, quasi-operational framework for supporting short-term climate predictions was constructed by enhancing key operational activities. This included the TOGA observing system, TOGA data centers, climate diagnostics bulletins, climate predictions, and others. GOALS needs to build further upon this existing framework.

The panel recognizes that it is important to coordinate GOALS activities with other operational and research activities of U.S. federal government agencies involved in supporting efforts to improve seasonal-to-interannual prediction. These include fundamental research into processes, observing systems, modeling, diagnostic analyses, data assimilation, data and information management, and applications. Several of these activities are also carried out at operational and research centers, as well as the observational programs of various agencies. Many are organized as part of the activity known as the Seasonal-to-interannual Climate Prediction Program (SCPP). The latter has grown out of the TOGA program and includes a number of components, as indicated in Figure 5-1 and 12-1. Many national, regional, and international programs strongly supported by the United States are involved in SCPP.

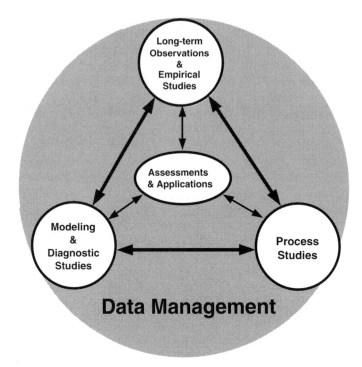

FIGURE 5-1 GOALS program architecture, showing interrelated components of the program. Data management is an activity in all components.

The International Research Institute for Seasonal-to-interannual Climate Prediction (IRI), has been established together with associated research and applications centers. The United States hosts the IRI as a U.S. contribution to GOALS/ CLIVAR. The National Centers for Environmental Prediction (NCEP), operated by the National Oceanic and Atmospheric Administration (NOAA) and the National Weather Service, cover climate prediction, environmental modeling, aviation prediction, and human dimensions aspects. The Climate Dynamics and Experimental Prediction (CDEP) program is a major element of the NOAA-OGP's (Office of Global Programs) climate program. NOAA's Pacific Marine Environmental Laboratory is primarily responsible for the development, deployment, and maintenance of the Tropical Atmosphere Ocean (TAO) buoy array in the Pacific Ocean.

The proposed GOALS research strategy is built on this rapidly developing infrastructure and the operational activities in various centers. Currently, the situation is markedly different from the research program established under TOGA, for which little infrastructure was in place during its early stages.

Linkages between the GOALS research strategy, and the operational activities mentioned above, should be seamless and symbiotic. GOALS will depend heavily on the various products and infrastructure provided by existing or planned operational efforts. On the other hand, GOALS research efforts will provide the knowledge, tools, and technical advances that will help accelerate the development of operational systems. To this end, the GOALS Panel strongly recommends that the following activities be initiated as a part of U.S. participation in international GOALS/CLIVAR:

1. Provide the basic research infrastructure to prospect for predictability within the global ocean–atmosphere–land–ice system in order to incrementally improve seasonal-to-interannual prediction skill.
2. Describe, analyze, and diagnose the physical processes that determine the variability in the observed fields and the spatial and temporal links among variables.
3. Develop pilot long-term monitoring systems of seasonal-to-interannual variability for the coupled ocean–atmosphere system.
4. Conduct process studies in select oceanic regions to improve the understanding of ocean–atmosphere coupling.
5. Develop, improve, and evaluate models of the coupled ocean–atmosphere–land system to be used for prediction.
6. Provide ongoing evaluation of climate monitoring systems and prediction products from IRI, NCEP, CDEP, and other operational systems.
7. Advise on the need for new products for multiple applications and help develop products addressing the applications and human dimensions aspects of the program.

The strategy proposed by the panel for U.S. participation in GOALS includes the appreciation that there is a critical need to establish mutual relationships with other national and international research programs involved in the study of short-term climate fluctuations. GOALS should explicitly take into account the activities of these other programs when developing its own strategies so as to minimize redundancies. For example, it is important to note that GOALS emphasizes process studies, empirical studies, and observations focused on ocean-atmosphere coupling. This emphasis is proposed because another component of the WCRP, GEWEX, is a complementary climate research program that focuses on the role of land in the coupled ocean–atmosphere–land system. Therefore, the panel recommends that GOALS should actively pursue collaborative research with GEWEX as the means of conducting process studies in select land areas. Close collaboration with GEWEX is also envisaged to develop pilot long-term monitoring systems covering the role of the land and land-surface processes in climate variability. Such joint activities are important to help achieve the global data coverage necessary to initialize and validate the complex coupled models developed by GOALS.

6

Long-Term Observations and Analyses

Sustained, accurate, long-term global observations of key variables in the climate system are essential to achieving the objectives of GOALS. A robust description of seasonal-to-interannual variations and the decadal modulation of these variations can be built up only from such an observational base. Diagnosis of the causes of these variations is impossible without long, accurate records. Continued global observations are also necessary to initialize and evaluate the global and regional coupled models envisaged under GOALS. These observations need to be assimilated into climate prediction models in much the same way that observations presently are assimilated into operational weather forecast models. The manner in which the observed fields are used to provide a description of climate and to develop hypotheses is illustrated in Figure 4-1. Observations to test specific hypotheses are gained through process studies. The conclusions reached from these data may be used to redefine the long-term observational strategy in order to describe the climate system more completely.

To be useful, measurements of seasonal-to-interannual climate variability must be consistent, continuous in duration, and of sufficient accuracy and resolution that the climate signals can be distinguished from instrumental noise and from high-frequency geophysical signals unrelated to climate variability. Because the phenomena being studied are generally of a large scale and influenced by processes in distant regions and because the intent is to improve climate predictions in locations remote from phenomena such as ENSO, long-term global coverage is essential for many aspects of GOALS. Thus, a principal component of the global monitoring of the state of the ocean, atmosphere, land, and ice variability must be a comprehensive satellite system. For other potentially predictable elements, where the intent is to use oceanic variability in one region to predict climate variability on a global basis, additional long-term observational needs may be specific to particular regions. As an example, SSTs and winds in the tropical Pacific are essential as initialization data for models used to predict El Niño and its impacts. ENSO model predictions can also be evaluated from surface air temperature and precipitation data at specific locations remote from the tropical Pacific.

During TOGA, a core in situ, regionally focused, long-term observing system was established in a research mode, for monitoring and forecasting seasonal-to-interannual variability of equatorial central Pacific SST (NRC, 1994b). This observing system provides a starting point for the in situ oceanic component of an observational system for GOALS. In conjunction with a comprehensive satellite observational program, the expansion of long-term oceanic observations under GOALS from the tropical Pacific to the tropical Indian and Atlantic Oceans and to higher latitudes is believed to be necessary to achieve additional improvements in seasonal-to-interannual predictability. This, of course, imposes significantly increased demands on observing systems as well as data management and data exchange systems.

An important consideration is that the viability of some existing operational systems is seriously threatened by internationally imposed restrictions on data exchange. An internationally agreed upon policy for the free exchange of and access to scientific data for research purposes is critical to the success of GOALS, and indeed all other research endeavors.

The deterioration of existing operational observing systems is of serious concern to the GOALS Panel. During TOGA, enhancements important to the monitoring and prediction of ENSO were implemented under the World Meteorological Organization (WMO) World Weather Watch (WWW) program. However the upper-air sounding system has continued to degrade since the First GARP Global Experiment (FGGE) in 1979. Currently, there are about a third fewer upper-air soundings per day because of reductions from two to one per day in many countries (including the United States), determined in some cases by the increasing cost of expendables. The viability of the upper-air sounding system is also threatened by external factors. While the quality of rawinsondes has improved in the past 15 years, the decommissioning of the Omega navigation system in 1997 has led to the loss of some wind soundings. Some systems that depend on Omega have transitioned to the use of the Global Positioning System, but rawinsondes using this technique are much more expensive. Notwithstanding the above situation, the greater long-term threat is the assigning of frequencies used for transmitting rawinsonde signals for the use of cellular phones. This means narrower available frequencies and, thus, more expensive transmitters. These issues are of concern because changes in the type(s) of upper-air rawinsondes or in their total number or frequency of deployment can alter the climate record as seen through analyses and reanalyses.

Of special importance are the existing operational observational and data exchange systems. For the atmosphere, the WWW should be maintained and improved. WWW, a part of the operational system developed for weather monitoring and prediction, is composed of the Global Observing System (GOS), Global Telecommunication System (GTS), and Global Data Processing System. The WWW involves national, regional, and world centers. The International Oceanographic Commission (IOC) of the United Nations Educational, Scientific, and Cultural Organization jointly with the WMO, coordinates the exchange of real-time and near-real-time oceanographic data via GTS. Delayed-mode exchange is managed directly by IOC's International Oceanographic Data Exchange system, which includes national, regional, and world centers.

Enhancements to observing systems, and the collection and exchange of data for climate purposes are being planned by the Global Climate Observing System (GCOS) (WMO, 1995a), which includes a significant array of oceanographic and land surface measurements. The full set of data required for oceanography is handled by the Global Ocean Observing System (GOOS). Both GCOS and GOOS rely heavily on existing operational systems and mechanisms.

In contrast to the systems already in place for atmospheric and oceanic data, the international infrastructure does not fully exist for the collection and exchange of data to characterize the land surface (and vegetation) and land processes pertaining to climate variability. The promotion and establishment of such an infrastructure are being discussed under the Global Terrestrial Observing System (GTOS) (WMO, 1995b). Efforts to coordinate the requirements stemming from GCOS, GOOS, and GTOS are being undertaken under the Integrated Global Observing Strategy (IGOS) in order to identify, without redundancy to the extent possible, key enhancements to the GOS that serves the multiple needs of these programs. The panel feels strongly that GOALS should collaborate and coordinate with all of these programs.

Another important consideration is that the data from operational observing systems (both space and surface based) suffer from various problems, among them accuracy, representativeness, calibration, and continuity over time, as well as inadequately documented changes in instrumentation, observing practices, and data processing procedures or algorithms. Although these problems are especially important for decade-to-century time scales, they are also important for seasonal-to-interannual time scale research such as represented by GOALS.

Although the problems summarized above are serious concerns, the greatest observational challenge for GOALS involves the ocean variables that typically have a very short length of record, poor spatial coverage, and poor resolution in time. The fundamental problem is that many of the in situ oceanic observations (e.g., the TOGA-TAO array and surface-drifter observations of SST), as well as many of the satellite ocean observations that are crucial to GOALS (e.g., altimetry), have been developed and maintained largely in research modes. At present, no federal agency has committed to a comprehensive operational ocean observing program specific to the needs of seasonal-to-interannual climate forecasting, and more generally, to climate research. One of the objectives of GOALS must, therefore, be to promote the transition of observations from a research framework to an operational framework in order to ensure long-term continuity of the observations. Note that the term operational in this context does not necessarily mean the transitioning of a system to an operating agency (within the United States, a NOAA line organization such as the NWS or the National Ocean Service), but rather a commitment to both the infrastructure necessary to sustain observations and to maintain the quality of the observations. An example is the TOGA observing system and, in particular, the TOGA-TAO array of buoys moored in the tropical Pacific. The latter is maintained in a quasi-operational mode by a multi-national research group spearheaded in the United States by the NOAA Pacific Marine Environmental Laboratory (PMEL).

While it is crucial to maintain and enhance existing observing systems, GOALS will inevitably have to rely on and utilize the observations planned by the next generation of Earth observing satellites and programs. Space-based satellite observations will play an important role in providing global data coverage for many of the key variables required by GOALS with a spatial and temporal resolution, which cannot be matched by in situ measurements alone. In addition to providing much of the information necessary for coupled climate models, satellite observations of seasonal-to-interannual variability of ocean, atmosphere, and land variables will be used to assess and refine the models. Importantly, these observations will also play a vital role in monitoring the interactive processes within the purview of GOALS on a global basis.

Space-based observations are already used as a powerful and complementary adjunct to in situ observing systems with the latter providing absolute calibration and satellites providing global coverage. Global scale fields of SST are best constructed through a well conceived combination of in situ and satellite data— a process that maximizes the advantages of satellite observations (e.g., global scale coverage and relatively accurate gradient information) while simultaneously compensating for deficiencies in in situ observing networks even though they provide direct absolute value measurements.

Continuous, uninterrupted satellite observations are required for at least the 15-year duration of GOALS, if the program is to achieve its objective of determining the predictability of climate (especially precipitation) over North America and other regions on seasonal-to-interannual time scales. Such long, continuous time series, exceeding the lifetime of individual satellite missions or instruments, are extremely difficult to acquire outside of operational programs. While many of the needed satellite-based observations of key atmospheric and solar forcing variables are already incorporated into U.S. and international operational meteorological satellite programs, few of the key oceanographic or land variables are observed from operational satellites. Lacking an operational framework for these important satellite observations, the research community has worked hard over the past decade to maintain the satellite observational systems in a research mode. These satellite systems have been supported financially by the National Aeronautics and Space Administration (NASA) in highly successful collaborations with international partners (primarily the French space agency CNES and the Japanese space agency NASDA).

Because of the long lead time required to build and launch dedicated satellites, firm commitments must be established in the near future to ensure continuous observations from these research satellites over the full duration of GOALS. The NASA Office of Earth Science sponsors the Earth Observing System (EOS) designed to observe atmospheric, oceanographic, and land variables of importance to climate change issues. The Office of Earth Science has provided a framework within which individual research-mode missions will be launched to provide high-quality measurements on a mission-by-mission basis with no assurance of continuity of measurements of any specific variable. The GOALS objectives require continuous time series through at least 2010. NASA's charter to demonstrate and extend technology and the budgetary pressures facing the agency raise questions about NASA's ability to continue to provide quasi-operational data for all of the important satellite instruments over the duration of GOALS.

Given the importance of the satellite measurements for studies of seasonal-to-interannual variability of the ocean and atmosphere, as well as the requirement for high-quality satellite data for decadal and longer time-scale climate change, steps should be initiated now to incorporate mature satellite technology into U.S. and international operational satellite programs. In the United States, present plans are to merge the Department of Defense and NOAA operational satellites into the National Polar-Orbiting Environmental Satellite System (NPOESS). The first of the NPOESS satellites is not planned for launch until 2007. Since the lead time to achieve operational status of a satellite sensor is a decade or longer, a coordinated long-range plan must be developed at the earliest possible opportunity in order to provide benefit to the latter stages of the GOALS program and any future research programs that evolve from GOALS.

KEY VARIABLES

The challenging combination of accuracy, spatial and temporal coverage and resolution, and duration requirements of the observational component of GOALS is compounded by the range of variables needed to address the scientific objectives of the program. At this early stage of GOALS, all of the relevant variables and processes have probably not yet been identified. Expanding on the TOGA experience and other experimental programs that took place during the last decade and a half as well as recommendations made in reports such as NRC (1994b) and WMO (1995a,b), a list has been compiled of priority variables considered essential for a sustained, long-term observational system to support seasonal-to-interannual climate variability research and prediction.

Table 6-1 summarizes the variables presently considered necessary for each of the three components of the coupled ocean–atmosphere–land system. The proposed list is divided into *state* variables and *external (or forcing)* variables. State variables are defined as variables necessary to monitor the "state" of the physical system or subsystem, while "external" or "forcing" variables are defined as those variables that define the degree of interaction between system components. Though from an Earth-system standpoint, the partitioning of the system into subsystems and into state and forcing variables is somewhat arbitrary, for many purposes, such categorization is conceptually useful. Within each group, variables are listed in approximate order of importance to the program. Some variables could appear under more than one category.

It is also worthy of mention that GOALS's data sampling requirements have not yet been completely established. The spatial and temporal resolution requirements for a global in situ observational system for research on short-term climate variability and prediction have not been reassessed in light of the knowledge gained from TOGA, the World Ocean Circulation Experiment (WOCE), and more than 5 years of TOPEX/POSEIDON altimeter data. Moreover, it is not apparent at present whether the expansion of moored ocean arrays such as TAO to mid-latitudes is the best way to obtain long-term, extratropical ocean observations. As mentioned above, observations from satellites presently in orbit or soon to be launched will, therefore, play a key role in achieving the global requirements of the observational component of GOALS, at least in the early stages of the program. However, satellites alone cannot meet the observational needs of GOALS. In situ observations such as the TOGA-TAO array, temperature and salinity profiles from buoyancy-adjustable subsurface floats, drifter observations, island-based tide gauge measurements, and Volunteer Observing Ships (VOS) observations of upper-ocean thermal structure and the overlying atmosphere also must be maintained to validate and calibrate satellite observations and to provide measurements of vertical structure and surface properties that cannot be obtained from satellite observations alone.

An important and overriding principle for the GOALS observing system is that individual components should not be developed in isolation which, historically, has often been the case. Priority should be placed on obtaining fields of variables especially useful for model initialization and evaluation. The specific platforms used to obtain the required measurements could be of secondary consideration, although they are important from the standpoint of implementing specific enhancements to existing observing technology. It should be clear from the list of key variables tentatively identified in Table 6-1 that a composite GOS comprising both space-based and in situ measurements is required by GOALS. In the design and implementation of the GOALS observing system, the stringent

TABLE 6-1 Key "State" and "External (or Forcing)" Variables Necessary for the GOALS Program. Variables Are Listed in Order of Priority.

Ocean

State Variables	External Variables
• upper ocean temperature	• wind stress
• upper ocean currents	• net surface solar radiation
• sea level	• downwelling longwave radiation
• upper ocean salinity	• surface air temperature
• optical absorption	• surface humidity
• sea ice extent, concentration, and thickness	• precipitation

Atmosphere

State Variables	External Variables
• wind structure	• sea surface temperature
• thermal structure	• net radiation at top of the atmosphere
• surface air temperature	• land surface variables (listed below)
• sea level pressure	
• water vapor structure	
• columnar water vapor and liquid water content	
• precipitation	
• cloud cover and height	

Land

State Variables	External Variables
• soil moisture	• precipitation
• snow cover and depth	• net surface longwave and shortwave radiation
• vegetation type, biomass, and vigor	• surface wind
• water runoff	• surface air temperature
• ground temperature	• surface humidity
	• evaporation
	• evapotranspiration

sampling requirements of some variables of interest must be recognized explicitly.

One of the major objectives of GOALS (especially its observational component) is to contribute to the development of IGOS being developed by the Committee on Environment and Natural Resources (CENR) of the White House Office of Science and Technology Policy (OSTP), within the framework of a comprehensive system of analysis to ensure that global-scale coverage of priority variables and calibrated products for research and applications can be provided to the scientific community. The GOALS Panel feels that the scientists participating in GOALS research programs should be involved actively in providing advice on various aspects of the long-term observing system for climate research and prediction as IGOS evolves.

Some of the issues surrounding observational requirements are likely to be resolvable through statistical analyses or numerical experimentation. Limitations in the quality and quantity of previous observations and in the physics of numerical models will almost certainly stall the satisfactory resolution of some questions. The research agenda of GOALS should be able to make a strong contribution to the future development of global climate observing systems by conducting appropriate process studies (see Section 7), and by making specific recommendations on enhancements needed for observing systems. Through process studies in the regions selected for focused experiments by GOALS, even more specific recommendations are likely for enhancements needed to regional observational networks. Of course, the research conducted under GOALS would specifically target the data and observing system needs to support seasonal-to-interannual climate prediction.

Notwithstanding the above, the observational responsibilities of GOALS should not be limited merely to the measurement of the variables for the duration of the GOALS program. As mentioned earlier, the seasonal-to-interannual focus of GOALS overlaps significantly with programs pertaining to both shorter and longer time scales. To be useful to other components of WCRP/CLIVAR and the USGCRP, long-term observations must be quality controlled, organized, and ultimately processed into gridded fields for use by scientists involved in GOALS research, as well as other programs such as DecCen and GEWEX. The GOALS program should, therefore, be active in the construction of data sets as well as in objective analysis and data assimilation. A close cooperation between the scientists involved in GOALS's model development effort and those involved in the development of observational data assimilation models is highly recommended by the GOALS Panel, because numerical models are an integral part of state-of-the-art assimilation procedures.

MEASUREMENTS AND OBSERVING SYSTEMS

The priority state and external observations identified in Table 6-1 will be obtained from a combination of current and new observing systems. In support of these, observation facilities need to be developed for transmitting the data to regional and international data centers.

The panel highlights several issues that need to be considered seriously in the implementation of the observing systems and networks supporting the scientific objectives of GOALS. They are also important for the programs complementary to GOALS such as GEWEX and DecCen.

Critical elements of existing observing systems should be maintained in order to obtain long, continuous, and consistent time series. In this regard, continuous and consistent in situ time series of key field variables should be given high priority. As noted previously, they are needed for model evaluation, empirical studies, ongoing calibration and validation of remotely sensed observations, and information that cannot be obtained from satellites. There may also be a need for new, long-term observational technologies to be developed and incorporated into operational systems following the findings of future process studies directed at improving models and their prediction skills. It is important that all measurements within observational programs are accompanied by sufficient "meta-data" in order to document changes in location, observing practices, satellite retrieval algorithms, instrumentation, and other characteristics such as exposure, space/time coverage, and accuracy. Where possible, use should be made of observational products from other programs, such as cloud and radiation products from the International Satellite Cloud Climatology Project (ISCCP) and precipitation estimates from the Global Precipitation Climatology Project (GPCP) and the Tropical Rainfall Measuring Mission (TRMM) (the NASA/NASDA satellite launched in November 1997 to measure tropical rainfall more accurately than ever before from a space-based system).

To help define and develop appropriate ocean, atmosphere, and land observing elements for incorporation into the GCOS, the GOALS Panel recommends that close coordination should be maintained with the DecCen component of CLIVAR and other WCRP programs such as GEWEX and ACSYS (Arctic Climate System Study). Among other activities, this coordination between programs should include the evaluation and prioritization of observed fields and the articulation of resolution and accuracy requirements.

Advantage should be taken of emerging new methods for real-time data transmission to enhance real-time data return and reduce the cost of transmission relative to the cost of making observations.

Also, the GOALS Panel considers several other issues to be of importance. These are the:

1. maintenance of the network of in situ sea-level observations;
2. calibration and validation of paleoclimate records relevant to seasonal-to-interannual climate variability;

3. calibration of global water vapor fields;
4. continuity of accurate satellite altimetry;
5. continuity of accurate satellite scatterometers; and
6. maintenance of the rawinsonde network throughout the tropics and the tropical Pacific, in particular.

In addition, there are other important issues that need to be addressed by the international community. These are the:

1. Reduction of costs of using the ARGOS system for transmission of remotely located in situ data; and
2. Reduction of the high cost of obtaining observations and analyzed fields from international data centers outside the United States.

For the coordination and implementation of the above issues, among other activities of importance to GOALS, the panel strongly recommends the establishment of a GOALS Interagency Project Office (discussed in Section 12).

DATA ANALYSIS AND ASSIMILATION; DATA REANALYSIS

GOALS requires the capacity to observe and use multi-disciplinary data covering the atmosphere, the oceans, the land surface, and snow and ice fields, among others. For observations from the combination of space-based and in situ platforms to be of most value to the modeling component of GOALS, they should be transformed to gridded fields. Consequently, methods to effectively combine the output from different measurement systems and platforms should continue to be developed. The combined use of altimeter sea-level data with thermal structure information to estimate the basin-scale field of upper-ocean thermal structure is an example of such a consolidation. New objective analysis and assimilation techniques should also be developed, especially for ocean and land surface variables to determine land surface processes, ocean–atmosphere interactive processes, and the hydrological cycle.

Historical data analysis warrants special mention. Because GOALS is concerned with seasonal and interannual time scales, it is important to compile a data record as long as possible so that many samples of seasonal and interannual phenomena may be examined. To this end the panel recommends that continued efforts should be made to improve the quality and volume of the historical data base through data archaeology and support for the development and evaluation of data sets. Typically, there is a need to correct and adjust historical data in various ways to compensate for changes in the manner in which measurements were made in the past. These data are used to determine the historical climate record, which provides estimates of the variability on interannual and longer time scales. Such variability includes the impact or influence of both natural and anthropogenic forcing components. Where possible, a systematic and periodic inter-comparison of historical data sets should be carried out.

The reanalysis of the data available from operational data assimilation schemes is an extremely important task for GOALS and currently an area of considerable activity. Much of the global gridded data fields used for research are based on past operational analyses techniques that are not ideal for climate studies. This is because the gridded fields contain inhomogeneities resulting from shifts in model and data assimilation/analysis methodology that have been tuned for operational weather forecasting purposes. These data fields have changed and will continue to change in time as forecast models and their corresponding data assimilation schemes are improved. For this reason efforts to reanalyze historical data should continue and should be applied to operational atmospheric analyses, all oceanographic data and analyses, and satellite-based products.

The panel underscores that reanalysis needs to be a periodic and ongoing effort. That is, the reanalysis system cannot be frozen in time, because data assimilation and modeling sophistication will certainly improve and change with time. Even the present reanalysis efforts, while highly commendable, have been found to contain several problems principally resulting from assimilation schemes and the state of models, both of which are still developing. In summary, the panel suggests that data reanalysis should become a routine activity that is repeated at regular intervals of time, conceivably every 5 to 10 years, in order to take advantage of the latest state-of-the-art systems and techniques available.

OBSERVATIONS WORKING GROUP

To carry out a thorough investigation of various issues dealing with long-term observations and analysis, the panel recommends that a GOALS Observation Working Group (GOWG) be established. This group should develop the scientific basis for GOALS observations. The activity is considered crucial because the scientific basis for observational requirements for climate studies and model evaluation is continually evolving. Furthermore, during the past decade and a half, our knowledge of the climate system has improved, resulting in changes in resolution and data type requirements. In addition, other forms of data, especially satellite, are now available. An important role of the GOWG will be to provide guidance in the prioritization of satellite-based measurements to be incorporated in the operational satellite program. Furthermore, an observations working group, working in conjunction with GEWEX, DecCen, GCOS, and the CLIVAR Upper Ocean Panel, would be well placed to make specific recommendations about future satellite instruments and products. Fuller community participation should be encouraged by the organization of one or more workshops to address data requirements for the support of GOALS objectives.

7

Process Studies

Process studies involve highly focused special experiments conducted to advance the understanding of specific physical processes. GOALS process studies should be enunciated or designed with the practical aim of improving seasonal-to-interannual climate predictions.

These special experiments often comprise a set of intensive measurements that are limited in duration and geographic extent. Usually, process studies employ an ensemble of techniques involving observations, empirical studies, and modeling to further the understanding of a process or processes. They may be in the form of experiments designed to test particular hypotheses or exploratory efforts to gather statistical information on specific processes or phenomena for use in developing and refining theoretical, mathematical, or conceptual constructions of the dynamics of various interactions between Earth system sub-components. Process studies can serve as test-beds for new observing platforms that may eventually contribute to the long-term observing system. They can improve our understanding of physical processes and help in the development of more realistic parameterization schemes in models, thereby leading to improved prediction skill. They can also provide critical tests of these models.

GOALS places priority on the improved understanding of processes leading to improvements in their representation or parameterization in coupled ocean–atmosphere–land models. Suggestions for specific process studies may emerge from the modeling community in order to address specific problems of model parameterization. The context of process studies in the GOALS program is shown in Figure 4-1.

Two basic categories of observational process studies can be defined:

Single process studies look at a particular process such as vertical mixing in the equatorial ocean, or near-equatorial atmospheric convection (e.g., as in EMEX: the Equatorial Mesoscale Experiment), or cloud-radiation feedback, and so on. They can, however, be quite complex in terms of the research and observations required for their investigation. Single process studies are relatively easy to plan, coordinate, and conduct because they often (but not always) fall within a single scientific discipline. Drawbacks of single process studies often include a lack of concurrent data and information regarding other processes and events occurring within the total system and thereby a lack of context relative to larger scales and processes.

Combined observational process studies are essentially two or more single process studies that are collocated or adjacent and are conducted contemporaneously. Combined process studies can provide a multivariate data set that has the critical mass needed to evaluate coupled model and satellite products. Such field programs can provide a facilitating framework for interdisciplinary studies, as well as a better context and supporting information than single process studies. Combined process studies are, however, usually more difficult to fund, organize, and conduct. They also require substantial collaboration among scientists from the different disciplines involved and, correspondingly, collaboration between the diverse funding sources, which are to this day oriented generally along disciplinary lines. The situation is, admittedly, improving rapidly with the USGCRP and federal agency emphasis on multi- and inter-disciplinary activities. The more complete picture that combined process studies provides makes them preferable to several isolated single process studies and experiments. Often, maximum benefit is obtained when single process studies are embedded within a larger more comprehensive combined process study or experiment. This strategy, endorsed by the panel, is also in accord with the thinking of other scientific groups and federal agencies in order to maximize the benefits from research experiments and their applications.

Process studies are expected to play an important role in formulating and testing hypotheses during GOALS, just as they have in earlier programs. For example, TOGA-COARE, the Coupled Ocean–Atmosphere Response Experiment (COARE), was developed as a part of TOGA to address the general problem of accurately quantifying air–sea fluxes of heat, moisture, and momentum in the warm-pool region of the western Pacific. Experimental modifications made to models covering the western tropical Pacific have led to the improved prediction of the global scale impacts of ENSO, illustrating the very tight coupling of the ocean and atmosphere in this region through the exceptional sensitivity of both the atmosphere and the ocean to changes in the other. COARE addressed a suite of processes that are important for coupled models but went further in conducting its observations in a key region for climate variations associated with ENSO and the global climate. The observations obtained in the experiment not only characterized the individual processes but also resulted in a benchmark data set in a critical region for testing coupled models and remote sensing algorithms. Thus, COARE was a combined process study that addressed both the prediction and the monitoring goals of TOGA while providing data for a wide range of physical processes relating to the heat balance of the tropical Pacific warm pool.

However, insufficient progress has been made during TOGA on achieving a better understanding of the more complex processes that affect and control the atmosphere–ocean exchange fluxes, which are intimately involved in the coupling of the ocean and atmosphere on larger space and time scales than the experimental region of TOGA-COARE. The mean state and the variability within the warm-pool region of the Pacific have been especially problematic for coupled models (and even stand-alone models) to simulate correctly. Thus, it is expected that new process studies will be proposed in GOALS.

EXAMPLES OF ONGOING AND PROPOSED PROCESS STUDIES

Learning from the experience of TOGA-COARE, the panel recommends that GOALS coordinate the design and implementation of essential process studies to advance our knowledge of several facets of GOALS simultaneously.

At present, proposals for particular process studies in GOALS are at an early stage of development. However, under the auspices of the PACS program, a number of process studies in the eastern Pacific Ocean dealing with ocean–atmosphere interaction, precipitation processes, and the equatorial boundary layer are already funded and currently under way. Ocean cruises took place in Spring 1997 to study the local modification of the equatorial ocean cold-tongue during the evolution of El Niño. An ocean cruise with weather radar instrumentation is occurring in Winter 1997/1998 and investigating changes in El Niño-related convection in the eastern Pacific Ocean. The Pilot Research Moored Array in the Tropical Atlantic (PIRATA) is being developed to monitor ocean–atmosphere interactions in the region of the Atlantic tropical SST anomaly dipole. PIRATA consists of an array of moored buoys that, among other objectives, will help determine a strategy for long-term observations in the equatorial Atlantic Ocean.

Several other international activities under the auspices of GOALS are also under way. The planning for the Asian–Australian monsoon system experiment (a GOALS regional focus) is proceeding with the implementation of the South China Sea Monsoon Experiment (SCSMEX). SCSMEX plans to study ocean–atmosphere interaction during the onset of the East Asian summer monsoon. A significant extension of the TOGA-TAO array into the equatorial Indian Ocean and northwest Pacific is being implemented by Japan. Process studies are in the planning stage to observe ocean–atmosphere interactions in the eastern Indian Ocean during intraseasonal variations of the monsoon. The expansion consists primarily of deep ocean moorings called the Triangle Trans-Ocean buoy network (TRITON) array. The TRITON moorings are modeled after the NOAA TAO moorings and contribute seamlessly to the ENSO observing system data. In order to improve the understanding of Atlantic modes of variability, NOAA, together with international partners, is expanding ocean/atmosphere observations into the tropical Atlantic in pilot mode. The Atlantic observing system is built from the same proven technologies used in the ENSO observing system, namely, deep ocean moorings (e.g., PIRATA), drifting buoys, tide gauges, and VOS. Additionally, a fifth network of autonomous profiling floats will be added to the system in order to measure better subsurface ocean currents, temperature, and salinity. The Atlantic observing system, together with the ENSO observing system, and the Indo-Pacific TRITON expansion form a significant contribution to the climate module of the GOOS. A Joint Air–Sea Monsoon Interaction Experiment (JASMINE) is being proposed for the Indian Ocean. JASMINE aims at studying ocean–atmosphere interaction in the eastern Indian Ocean during the intraseasonal transition of the South Asian Monsoon. In addition, extended observations of the ocean–atmosphere interaction within the Indonesian Archipelago and the Indonesian Throughflow are planned.

Complementary land process studies are being proposed under the framework of GOALS, but their implementation will be coordinated with and undertaken by GEWEX. An example is the GEWEX Asian Monsoon Experiment (GAME) under which land surface processes are being investigated in a variety of climatic regimes ranging from tropical forests to tundra.

GUIDELINES FOR THE SELECTION OF PROCESS STUDIES

For the further development of process studies beyond those mentioned above, the panel suggests that the following criteria and activities be kept in mind:

1. Whenever feasible, single process experiments needed for GOALS should be coordinated and combined, although it is realized that individual pilot studies may be necessary at first.

2. Domains for GOALS process experiments should be chosen to include oceanic regions where SST or heat content variations are likely to provide a relevant atmospheric predictive capability.

3. Process experiments for GOALS should be designed to provide a critical mass of ocean and atmosphere multivariate data from in situ and remotely sensed observations to support adequate future model experimentation. They should also provide a basis for the development and evaluation of relevant remote sensing techniques.

4. Combined process experiments should be undertaken in key regions for seasonal-to-interannual climate variability, as demonstrated, for example, by model sensitivity studies.

5. Process studies should be designed for use as test-beds for new observing

techniques that provide enhanced monitoring of priority variables in limited but important regions.

6. Long-term observations should be used to provide the context for limited-duration intensive process studies.

7. Workshops should be held to enable scientists to develop consortium proposals for necessary process studies.

The panel recommends that GOALS process experiments should be coordinated either with GEWEX process studies, such as GAME and the GEWEX Continental-Scale International Project (GCIP), or with DecCen process studies. Close coordination is considered necessary with climate-related U.S. and international programs to achieve mutual objectives. This is important for producing a critical mass of observations as well as for sharing the funding of process studies.

The TOGA investment in COARE and other relevant process studies should be leveraged by continuing support of analysis and modeling activities that use observations from associated field programs.

8

Empirical and Diagnostic Studies

Empirical studies and diagnostic analyses are essential elements of GOALS. Empirical studies describe phenomenological relationships within and among fields of different variables, both temporally and spatially. Diagnostic studies include quantitative analyses of processes and budgets and extend beyond empirical relationships. Together they lead towards an ability to describe and understand what is occurring in nature and how such knowledge can be used to improve models.

Empirical studies proposed under GOALS should encompass a broad spectrum of activities ranging from investigations that search for empirical relationships with predictive value to investigations intended to produce a better understanding of processes leading to improved parameterizations for prediction models. In keeping with the overall objectives of GOALS, priority is accorded to diagnostics and systems studies pertaining to relationships and processes that influence the seasonal-to-interannual predictability of climate variations, first over North America and then globally.

The role of diagnostic and empirical studies in the composite structure of GOALS is shown in Figure 4-1.

The basic strategy for GOALS calls for an extension of the TOGA emphasis on the Pacific to the global tropics in order to encompass the major heat sources and sinks associated with the tropical oceans and major land masses. This expanded scope should include the interaction between, for example, the monsoons and ENSO, and between the tropics and the mid-latitudes. The probable interaction of phenomena on a wide range of space and time scales will require very careful empirical and diagnostic analyses.

STRUCTURES AND INTERRELATIONSHIPS

It is possible that hitherto unknown relationships between the major components of the climate system will add to current estimates of its predictability. Such relationships could suggest physical mechanisms that influence the system and indicate possible sources of predictability that can be exploited. Possible candidates are the interaction between Intraseasonal Oscillations and ENSO, and between ENSO and the Tropical Biennial Oscillation, and the Quasi-Biennial Oscillation.

DIAGNOSTICS

Diagnostic studies are excellent tools for posing and testing hypotheses as well as providing basic information about budgets, fluxes, and other parameters not measured directly. The climate system is so complex and interactive that a wide range of diagnostic studies will be needed to investigate and test hypotheses on the underpinning physical processes governing various interrelationships. Examples of other specific products obtainable from diagnostic studies include estimates of budgets of atmospheric mass and moisture and of atmospheric and oceanic heat; fresh water and momentum fluxes and budgets; and the movement and evolution of ocean heat content, and the surface heat balance. The new ocean data collected under GOALS, will enable the determination of the role of planetary-scale ocean waves (including the "delayed oscillator" mechanism) in seasonal-to-interannual climate variability.

FORCINGS—RESPONSE AND FEEDBACKS

Central to the understanding of seasonal-to-interannual climate variations is the response of the atmosphere and oceans to external forcings. Also important are linkages between local and remote locations and the impact of various feedback mechanisms and processes between components of the ocean–atmosphere–land system. Several aspects of forcings and feedbacks are proposed below for further study.

1. The spatial and temporal distribution of atmospheric forcing should be determined, with the aim of finding which aspects are most important for seasonal-to-interannual variability and why such forcings change. Among the most easily observable symptoms of a forcing change are changes in the convection zones in the tropics. These changes provide two of the most important feedbacks in the atmospheric system—latent heating and cloud–radiative interaction—and

respond to changes in boundary conditions both locally and remotely. Perhaps the most important hypothesis to be tested is that changes in convection zones provide the major pathway connecting seasonal-to-interannual atmospheric variability with forcing by the more slowly evolving parts of the climate system. Convection anomalies are of interest both because of the importance of precipitation to human activities and because of teleconnections to other parts of the atmosphere. Thus, identification of mechanisms by which boundary condition forcing produces such changes in convection zones is a priority.

2. The response of the atmosphere and oceans to local and remote atmospheric forcings, and remote responses in the oceans arising from atmospheric teleconnections, should be assessed empirically. Many questions concerning wave propagation, interaction of quasi-stationary waves with transients, and so forth, can be addressed empirically as well as with modeling studies. These questions are important because they lead to a physical understanding of the way predictable features of the tropics are projected to other regions.

3. The impacts of local and remote forcing on the hydrological cycle (especially over land) should be determined, and feedbacks to the atmosphere through changes in ground hydrology and other land surface processes should be estimated. The role of snow cover and soil moisture over land in inducing anomalies in large-scale circulations should likewise be determined. It is important to assess the extent to which land surface anomalies are stochastic or part of long-term fluctuations.

4. The importance of other feedbacks on the coupled system, such as changes in storm tracks, interactions with sea ice, and effects of extratropical SST anomalies, should be assessed.

PREDICTABILITY

A primary focus of GOALS is the investigation of predictability with empirical and diagnostic studies. To this end it is recommended that analogues be used to determine the rate of separation of initially similar states of the coupled system. But because of the limited time span of historical data record, innovative approaches are needed to address the long time scales of interest in GOALS. Empirical searches should also be made for regular behavior; very-low-frequency phenomena; and spatial, temporal, or mechanistic links between elements of the coupled system, in order to identify elements or features that may lead to improvements in prediction skill.

9

Modeling

Modeling is an important means of understanding the physical structure of the coupled climate system and assessing its predictability. Modeling serves as a means of synthesizing the long-term observations, process studies, and empirical and diagnostic studies of GOALS and focusing them on the program's central objective—the development of skillful prediction of seasonal-to-interannual variations of the coupled ocean–atmosphere–land system. Therefore, a comprehensive modeling program is an essential component of GOALS.

To realize the full potential of models as an aid in achieving the scientific objectives of GOALS, a complete hierarchy of models needs to be developed and applied to various problems. The panel feels that global General Circulation Models (GCMs) and intermediate-scale models are the most appropriate models for quantitative climate simulations and prediction experiments. Intermediate and mechanistic models are recommended to isolate fundamental processes. Empirical models constructed on the basis of relationships found in the historical record are thought to be best used in the explorations of predictability and to act as a benchmark with which the performance of other models can be compared.

The panel suggests that the modeling efforts under TOGA, which dealt with time and space scales similar to those addressed by GOALS, can serve as examples of the range of activities envisioned for GOALS and as the starting point for modeling in the new program. It is recalled that skillful forecasts by empirical models identified some of the predictable components of the climate system. Later, highly simplified models disclosed the fundamental processes that influence the components of the coupled system and provided insight into its predictability. These processes were studied more deeply with sophisticated stand-alone component models and in simplified coupled ocean–atmosphere models. The simplified coupled models were used with moderate success in prediction. They indicated the robustness of the predictability of the coupled system and demonstrated the utility of deterministic models as climate prediction tools. However, the simplified coupled models used were usually of limited geographic extent and depended on mimicking or approximating the physical processes involved in ENSO. The only product of many simplified models was a forecast of SST in the tropical Pacific Ocean region. They did not take into account other large-scale circulation features (e.g., the monsoons), nor did they provide attendant forecasts of precipitation, insolation, or circulation either local to the Pacific basin or remote from it.

To address the scientific objectives of GOALS and seek forecasts for the global domain, more complete modeling is thought to be necessary. It is especially important to develop models that have the ability to simulate the annual cycle and interannual variability and to predict seasonal-to-interannual variations of circulation and rainfall, two significant benchmarks for GOALS modeling.

In addition to continuing the effort to understand better the coupled system and develop more comprehensive models of it, modeling efforts in GOALS should also be applied to tasks such as determination of the predictability of the coupled system, experimental prediction, data assimilation, and observing system design.

The GOALS modeling process and its relationships to the other elements of GOALS are described in Figure 4-1.

Five principal types of activities are recommended for the GOALS modeling components. These are:

1. model development;
2. predictability and sensitivity studies;
3. experimental prediction;
4. development of data assimilation systems; and
5. observing system simulation.

MODEL DEVELOPMENT

In general terms, the panel believes that improvements are needed in all of the component models of the climate system in order to increase prediction skill in the seasonal-to-interannual time frame. In particular, the interface fluxes of various quantities such as heat, moisture, and momentum need to be represented more accurately in the models.

Atmospheric General Circulation Models need to be improved to a stage where, when driven with prescribed observed SST, they simulate realistically the observed annual cycle and interannual variability of the surface wind stress and heat flux, as well as the global atmospheric circulation and rainfall. Better atmosphere and ocean models are expected to evolve from a systematic program of diagnosing and evaluating models with improved parameterizations. Similarly, Oceanic General Circulation Models need to be improved to a stage where, with prescribed surface stresses and heat fluxes, they simulate realistically the observed annual cycle and interannual variability of SST, upwelling, upper-ocean heat content, convection, and subduction. The improvement of Land Surface Process Models (LSPMs) to a stage where they represent adequately the interaction between the land surface and vegetation and the atmosphere is necessary. Serious consideration must be given to the incorporation of a dynamic land-surface component to existing models such that the characteristics of the land surface and the state of the biosphere evolve with changes in atmospheric circulation, clouds, precipitation, radiation, and so forth. On the large scale, LSPMs should also be constructed to account for the significant time lags between winter snow or ice and their melt in spring, leading to ground water recharge in spring. The coupling of land surface processes with the oceans through river outflow should also be examined. On shorter time scales, especially in the tropics, LSPMs should be able to characterize moisture recycling processes. Evapotranspiration through the vegetation canopy and albedo changes and feedbacks are important processes that have to be modeled better.

In addition to improving the individual component models of the climate system, the panel recommends an equally strong effort in coupling the component models together so that they better approximate the natural system. In this connection, the composite models should be tested by assessing their ability to simulate and predict key parameters and variables. Significant improvements are considered necessary in coupled ocean–atmosphere–land models so that they can realistically simulate the observed annual climate cycle and its higher frequency components, as well as the statistics of the observed interannual variability of tropical SST, land surface temperature, the global atmospheric circulation, and rainfall. These models will need to take into account land-surface hydrology processes, a requirement best met by close cooperation with GEWEX modeling groups working on the parameterizations of these processes.

To minimize the complications that arise when linking component models together, the panel recommends that coupled models should be developed with modular "plug compatibility." This would enable a broader participation of the scientific community in the development of models and also enable the exchange of various parameterization schemes between models and between scientists. If possible, protocols for "plug-compatibility" standards should be developed.

Though somewhat simpler than GCMs, models of *intermediate complexity* should also be constructed. These models can be compared quantitatively to observations for specific applications. Their uses include isolating fundamental processes, sensitivity studies, exploratory predictions, and ensemble predictions where computational costs are important. For GOALS applications, greater attention to land processes, physical parameterizations, and consistent energy budgets and atmospheric eddy fluxes is highly recommended. Simpler models should be developed for the study of specific processes thought to be important in nature. By making it possible to study processes in a controlled setting, such models can lead to the identification and characterization of those mechanisms that affect seasonal-to-interannual prediction.

With regard to applications and human dimensions, the panel recommends that *auxiliary models* be designed to predict societally important quantities not routinely produced by seasonal-to-interannual climate forecast models. These models can be based on both empirical and physical techniques and could involve predictions of quantities including regional rainfall and storm track activity that are not predicted by intermediate models or, for example, the likelihood of tropical storms or extreme events that are not predicted by coupled GCMs. Auxiliary models include those used to make projections of agricultural yield, water availability, fish productivity, energy demand, economic impact, and so on.

Improved strategies should be developed for nesting high-resolution regional models and global climate models to infer detailed structures of regional climate anomalies forced by global boundary conditions predicted by coupled ocean–land–atmosphere models. Furthermore, the systematic and periodic intercomparison of models should be continued in order that the physical sciences and the user communities have available an ongoing assessment of the status of models.

PREDICTABILITY AND SENSITIVITY STUDIES

The determination of the limits of predictability on the seasonal-to-interannual time scale includes not only traditional numerical experiments designed to examine global assessments of the limits of predictability, but also investigations of the geographical, seasonal, regime, and field dependence of model predictions. Individual component and coupled models of a broad range of complexities will be required to contribute to this task. Whether ENSO predictability is sensitive to high-frequency atmospheric forcing needs to be determined as well as the degree to which ENSO irregularity and limits to predictability might result from chaotic behavior in the slow components of the coupled system.

The panel recommends that the dynamic and physical mechanisms that contribute to internal and forced variability on seasonal-to-interannual time scales should be investigated. The aim of the investigation is to enhance our understanding of the causes of internal and forced variability so that results from comprehensive models and empirical studies can be interpreted. For example, what is the role of extratropical SST anomalies in climate variability? What is the predictability of the annual statistics of tropical storms? In particular, quantitative estimates need to be made of the range of time for which initial conditions of the atmosphere are important. This determination is crucial for the predictability of midlatitude variations. Another example is the determination of how the probability distributions of planetary waves respond to tropical heating anomalies, interactions with the annual cycle, ENSO cycles, monsoon variability, and climate change.

Other activities recommended by the panel include experiments to determine the contributions of the upper-ocean heat content (tropical and extratropical) and land surface moisture and heat content to the sources of memory for seasonal-to-interannual variability. Also recommended are assessments of the relative sensitivity of climate system response (and its time evolution) to uncertainty in initial conditions and boundary conditions.

EXPERIMENTAL PREDICTION

Central to GOALS is the development and implementation of experimental prediction projects directed at establishing the seasonal-to-interannual prediction capabilities of models, and to enhance the information content of their forecasts in regard to modeling applications. For example, it is important to determine the skill of state-of-the-art models in hindcasting past variations of the climate system. This should include an evaluation of component models with specified boundary conditions in addition to the evaluation of the performance of coupled models. Techniques such as ensemble forecasting could be utilized effectively for estimating the probability and distribution of future climate system states. The panel also feels that improved statistical and dynamic techniques need to be developed for assessing and predicting the skill of model forecasts. To this end, the panel recommends the establishment of standardized procedures and protocols for the ongoing evaluation of experimental forecasts and the comparison of these forecasts with empirical/statistical prediction methods.

Of particular importance is the identification of regions, circulation features, or phenomena that have above-average predictability on seasonal time scales. Moreover, real-time predictions of climate variations using different models and techniques should be carried out in addition to retrospective and research mode model experiments.

DEVELOPMENT OF DATA ASSIMILATION SYSTEMS

Data assimilation, which is also discussed in Chapter 6 under "Data Analysis and Assimilation; Data Reanalysis," is an activity that cuts across more than one GOALS element. The improvement of data assimilation methods, especially for the ocean and land components (and the coupled system), and demonstration of the usefulness of these techniques for experimental prediction are important aims of GOALS research. The panel emphasizes that four-dimensional data assimilation is essential for defining mutually consistent states of the atmosphere, the ocean, and the land surface as required by global coupled models. This should also enhance the predictability of the coupled system. The large cost of GCM data assimilation implies that this activity needs to be carefully coordinated with the operational centers, including NCEP, the IRI, and NASA, among others.

OBSERVING SYSTEM SIMULATION

Models are useful for estimating the utility of particular measurements. To the extent that models accurately simulate the statistics of the full spectrum of energetic space and time scales that occur in nature, they can be used to set resolution requirements that will avoid undersampling and aliasing. Ocean models are just approaching the point where they can contribute such assessments of observations for climate research purposes, but the empirical approach, where deliberate attempts to oversample are made in order to check for aliasing, is still needed. Atmospheric models, having been tested more thoroughly over the years, are currently more reliable for this purpose. Modeling experiments can also be used to determine the minimum observational system needed to induce data assimilation systems to generate the observed circulation fields to within acceptable tolerances. Scenarios for observations during the Global Weather Experiment (GWE; also known as FGGE) were evaluated with atmospheric general circulation models, for example. A unique requirement for GOALS is to assess elements of the observing system in regard to their utility for supporting predictions on seasonal-to-interannual time scales. The most efficient mix of observing techniques that will deliver the required accuracy and space–time resolution will be sought. The corresponding need for observing system simulation experiments with both component models and coupled models is a particular challenge to GOALS.

MODELING WORKING GROUP

The panel strongly recommends the establishment of a GOALS Modeling Working Group (GMWG), in order to involve a broader cross-section of the scientific community and to encourage and coordinate a program for model intercomparison, climate system simulations, and experimental predictions. The GMWG should encourage a multi-model approach and, in particular, facilitate the interchange of fully documented and fully accessible models and data sets. The GMWG should also develop protocols for plug compatibility between models and standardized procedures for the ongoing evaluation of model simulations and experimental predictions. Other activities should include plans and estimates for computing and communication needs and the coordination of specialized numerical experiments. Under CLIVAR, the international Numerical Experimentation Group (NEG) is devoted to modeling issues on seasonal-to-interannual time scales. The GMWG, in consolidating the efforts of a broad group of U.S. modelers, should coordinate closely with CLIVAR/NEG to optimize international modeling efforts and cooperation.

10

Applications and Human Dimensions

One of the fundamental objectives of GOALS is to assist in the development of an integrated climate prediction system for a broad range of applications on seasonal-to-interannual time scales. To reach that goal, the panel recommends that GOALS should encourage research activities in applications and human dimensions. The *applications and human dimensions* component of GOALS represents those activities that need to be undertaken to assess the implications of seasonal-to-interannual climate variability for human and natural systems. Human activities encompass, for instance, social, industrial, agricultural, hydrological, fishery, commodity, and various other economic activities. Importantly, as prediction skill improves as expected through the implementation of GOALS, it is crucial that the predictive information provided be in a form that is useful for practical applications in social, economic, industrial, and other resource management activities. In the past, the physical scientific community has grappled with this problem, albeit with some lack of success on account of the rather substantial gaps between the needs of the user for information and the products generated by forecast models. Often, there has been a significant deficiency in the lead time provided by forecast guidance and in forecast accuracy. These shortcomings will, hopefully, be bridged through the implementation of GOALS.

The activities envisaged under this component of GOALS will, by definition, be multidisciplinary and cross-sectoral. The GOALS Panel recognizes that the interfaces between the various communities involved in research and applications (particularly human dimensions) have not been explored before under any other physical sciences research program. As such, this aspect of GOALS poses new challenges to the organization of the implementation of the program, hence the reason for the panel to highlight this element as a specific objective of GOALS.

Close collaboration and coordination will be needed between physical scientists, and their colleagues in the biological sciences, social science, economics, hydrology, public policy, and others who participate in programs pertaining to the human dimensions of global change. This involves, for example, the applications community (those interested in variations in rainfall and temperature as they affect agriculture, fisheries, and water resources), the social science community (those interested in understanding impacts on human systems and institutional issues associated with the use of forecast information), and the physical sciences community (those involved in producing climate information and the development of forecast models and operational climate predictions). For example, for economists to determine the economic value of increased forecast skill in a particular region, the physical science community should identify appropriate regional climate variables, choose appropriate forecast representations related to those variables, and characterize the skill level and limitations associated with the probabilistic forecasts generated, typically, by models. Conversely, for physical scientists to develop predictive information products tailored for specific regions or sectors, the social science and economics research communities need to provide insights into the information necessary to assess the impacts on various sectors, the nature of potential users, and their specific forecast product needs.

A two-pronged approach that is well coordinated is considered necessary with a range of activities to be undertaken by the two communities. At this stage, it is felt premature to specify, with any great precision, the exact types of projects that should be considered. There are several issues involved in the interaction between the different scientific communities, including rather vast differences in terminology and importantly the specification of requirements for observations and model output products. These clarifications should be developed via an ongoing dialogue between the communities under this component of GOALS. Generally, two classes of efforts appear to be needed: (1) research into human dimensions and other applications such as those supported by the USGCRP and the International Human Dimensions of Global Change Programme (IHDP) with a focus on the seasonal-to-interannual time scale and (2) a specific effort by the physical science community to provide practical applications of physical science research that take into account needs identified by the human dimensions community.

The GOALS Panel has deliberated carefully on how to proceed in this somewhat unexplored area. In the past, the research and applications products developed were primarily of interest only to the physical science community. Other applications sectors had to interpret these products the best they could. Moreover, the requirements of the applications and human dimensions sectors were not integrated into the planning of research experiments in climate prediction.

The panel feels strongly that this situation needs to be rectified in order that maximum benefit is obtained from improvements in studies on climate variability and seasonal-to-interannual prediction. The panel further recognizes that this objective of GOALS cannot be realized without the direct participation of and interaction with these other communities.

The proposed interaction between physical sciences and the user communities is illustrated in Figure 4-1.

DETERMINING VULNERABILITY AND CHARACTERIZING IMPACTS

A considerable effort is needed for information to be gathered and analyses to be performed by applications researchers to determine the economic sectors within a region considered to be most vulnerable to climate variability and climate change. A further dimension in complexity is introduced when attempts are made to quantify and characterize the impacts of climate variability. Good examples of the problems associated with determining vulnerability and the characterization of social and economic impacts are the periodic assessments produced by the Intergovernmental Panel on Climate Change (IPCC) under the auspices of the United Nations (IPCC, 1990, 1996a). While the time scale of interest to the IPCC is long-term climate change, and that specifically induced by anthropogenic activity, the broad subject of applications and impact assessments contains the very same ingredients that will need to be addressed by applications in the seasonal-to-interannual time scale. To an extent, the shorter time scale injects a more serious challenge because of its relatively rapid verifiability. In fact, the credibility and accuracy of the forecast products, and the manner in which they are put to use, is likely to determine their acceptance by the broader applications communities where decisions are made on a daily basis that have a direct impact on finances and resource allocations.

The activities recommended by the panel to be carried out to investigate vulnerability and impacts of seasonal-to-interannual variability include determination of the following:

1. economic sectors in a given region most susceptible to seasonal-to-interannual climate variability;

2. manner in which societies and social institutions have adapted or responded to climate variability in the past, as well as identification of the insights that historical experience can provide for new forecasting systems;

3. sectors of society that could most benefit from improved forecasts;

4. impact of seasonal-to-interannual variability on water resources and the implications of this variability for flood control and conservation decisions;

5. impacts of different crop productions associated with seasonal-to-interannual climate variability on other commodities and trade;

6. impacts of climate variations on agricultural productivity and on the national economies of various countries;

7. human health impacts of ENSO and monsoon as well as their impact on, for example, natural resource management and socioeconomic activity;

8. impacts of ENSO and monsoon variability and other predictable elements of the climate system on precipitation, wind, storm tracks, hurricanes, floods, and so forth in key regions of the globe; and

9. impacts of other predictable modes of seasonal-to-interannual climate variations on crop production, forestry resources, fisheries, water resources, and others.

The need for "regional specificity" presents a particular challenge for GOALS. Both the applications and the social science communities require regional information that is as detailed as possible. Yet predictability inherently becomes less accurate when applied to smaller scales where the natural variability is larger. To address this issue, the panel recommends that the GOALS community should:

1. make special efforts to enhance regional forecasts where predictability has been or can be demonstrated and develop prediction products tailored for specific regions and sectors;

2. develop techniques to characterize the uncertainties associated with those regions where large variability is unrelated or only partially related to predictable large-scale forcing;

3. undertake investigations into the limits of regional predictability; and

4. develop predictive capabilities of the environmental quantities most useful for applications in a particular region.

CHARACTERIZATION OF INFORMATION NEEDS

The panel recognizes that there are several important issues currently inhibiting the interaction between the physical science community and the applications communities to whom forecast products are made available. Some result from a lack of understanding of the needs of various applications sectors or due to limitations in current forecasting skill, while others stem from difficulties in transforming the typical probabilistic forecasts into informed decision making. One major problem, namely the lack of lead time provided in forecasts, should be substantially alleviated by the implementation of GOALS, given its primary focus is the improvement of climate predictions in the seasonal-to-interannual time scale. Nevertheless, improved and more specific requirements for information are required. To this end, the GOALS Panel recommends that the applications sector and social scientists include the following in their activities:

1. an assessment of decision-making frameworks and mechanisms currently in place and how they can be adjusted to utilize forecast information more effectively;

2. the evaluation of the value of improved forecasts and the cost impact of forecast failures in various applications sectors; and

3. an assessment of the role of institutional responses to the impacts of climate variability and climate change and the options available for adjustments or adaptation.

Similarly, the panel recommends that the physical science community undertake the following activities:

1. review the information and climate prediction products that can be provided in the seasonal-to-interannual time scale today, and suggest steps that can be taken to address currently unmet requirements; and

2. work with the applications community to determine those variables that, when predicted, will have optimal impact.

ASSESSMENT TECHNIQUES AND COMMUNICATIONS

In the past, environmental assessments have been conducted primarily within individual sectors, and usually when confronted with a specific problem or issue. Often, they are retrospective and provide little guidance on how management decisions are to be made in the future, particularly in those regions susceptible to seasonal-to-interannual climate variability. The panel feels that this is an area that requires serious attention as a part of the implementation of GOALS. Even today, there are several examples of where seasonal-to-interannual prediction information can be used to help decision making in agriculture, water resource management, the distribution of fuels for energy, and the mitigation or prevention of vector-borne disease outbreaks. Most of these examples derive from the currently understood patterns of ENSO impact on global and regional temperature and precipitation distributions.

The expanded scope of the research that will be conducted under GOALS should improve significantly the accuracy and lead time of climate predictions, and their potential impact on various applications sectors. The panel also feels that a much broader multi-sectoral and multi-disciplinary approach is needed to develop techniques in integrated assessments. Furthermore, there is a need to facilitate the communication of information and products to and between the different scientific and user communities.

To pursue the above, the panel recommends that the applications and social science communities include the following activities in their programs under GOALS:

1. an assessment of how probabilistic forecasts of climate variations could be incorporated in resource management decision making (i.e., decision making relative to the uncertainty of the forecast); and

2. the development and refinement of improved assessment techniques that characterize and integrate the implications of seasonal-to-interannual climate variability.

Corresponding to the above, the physical science community should investigate how prediction information, albeit with uncertainty, can be presented in terms most useful for decision makers.

INSTITUTIONAL MECHANISMS

The execution of the applications and human dimensions aspect of GOALS will need institutional mechanisms involving national and international programs directed at transforming and using the results of scientific research. The panel recommends that close and formal links be established with existing and future institutional structures involved in such activities. The IRI, the Inter-American Institute, the WMO Climate Information and Prediction System (CLIPS), the IHDP, and the SCPP are highlighted as important initial programs for this activity. There will most likely be others as GOALS is implemented.

To facilitate the above activities, close coordination is recommended between the GOALS Panel, and its working groups, and the NRC's recently formed Committee on the Human Dimensions of Global Change.

11

Data Management

Data management is identified by the panel as a critical cross-cutting element of GOALS. This activity refers not only to the management of observational data obtained from various operational and research observing systems and networks, but also to data output (and products) from empirical and diagnostic studies, data assimilation systems, and climate prediction models as well as analysis products and information from the applications and human dimensions sectors involved in GOALS.

Each of the elements of GOALS will generate large multivariate data sets that are potentially useful to a sizable community of researchers. In developing a management strategy for these data sets, the panel recommends that the following principles be taken into consideration to maximize the utility of critical data and information resources:

1. free and open access to all GOALS observations, model products, and model code. This is a basic tenet of the GOALS program.

2. a distributed data management system. This model provides the most efficient access to existing products and encourages the incorporation of new products.

3. The development and maintenance of a GOALS data catalog will be a critical step towards the success of the distributed data base system proposed for GOALS.

4. real-time data transmission to the extent possible or feasible. Overcoming the technical and financial barriers that limit real-time transmission of GOALS data is a high priority of the GOALS program.

5. Adopt the FGGE classification system for data, data analyses, and information products, namely: Level 1 designating raw data, Level 2 for quality-controlled data, Level 3 for analyzed fields, and Level 4 for model output and data-information products.

6. Metadata and documentation of data processing techniques and dataset evaluation is essential and should be readily available.

Data management structures already exist under the auspices of WCRP on the federal agencies that launch and operate satellites and that routinely process, archive, and distribute the satellite data. The panel recommends that GOALS work with the USGCRP, the World Data Centers, federal satellite agencies, and WCRP data management structures to minimize redundancies and costs and to maximize data distribution in order to promote efficiency in data availability for research and applications.

The management of data from GOALS process studies will require special attention. It is important for the respective project scientists and managers to articulate data management requirements, develop a cohesive plan, and obtain funding and institutional commitments for specialized data management needs.

Each data set should be accompanied by a comprehensive meta-data file that documents information about the data, how they were collected, quality controlled, and processed, and their characteristics, resolution, and so forth. In addition, special attention should be paid to:

1. changes in observation times;
2. changes in the exposure of instruments, station environment, and the effects of urbanization;
3. changes in spatial distribution and aerial coverage; and
4. changes in methods of processing and analyzing observations.

A comprehensive data management program should also involve the resurrection of existing data sets (i.e., an effort aimed at the retrospective collection/compilation of missing data in the historical data archives). It is underscored that the total extent and magnitude of the observations made over the past 100 years far exceeds the data currently available for research purposes. While there have been commendable efforts to resurrect past observations from various national and international archives, such efforts need to be continued with adequate funding. These data are particularly important for the investigation of seasonal-to-interannual climate variability. They are even more important for studies on longer time scales. This task is usually underestimated by funding agencies.

Several of the above points and issues that have been proposed for establishment by the panel (NRC, 1995) and reiterated throughout this document, are elaborated in Bits of Power: Issues in Global Access to Scientific Data (NRC, 1997).

12

Coordination and Implementation Considerations

The panel considers coordination with other programs, both nationally and internationally, to be a crucial aspect for the successful implementation of GOALS because of the overlapping time scales of interest in climate-related issues and research, and importantly the interfaces that need to be built between the various scientific communities involved. Conceivably, for the first time, all of the physical scientific disciplines are involved. Adding complexity, the social and human dimensions disciplines (endorsed by the USGCRP) are also proposed as important partners. If these objectives are to be achieved by GOALS, extensive coordination will be necessary.

INTERNAL LINKAGES

The panel considers it extremely important for internal linkages to be established between the six elements of GOALS. That is, each element of GOALS must be cognizant of activities in all other elements of the program. In particular, there is a need to forge strong links among those making observations, the research and applications users of the observations, those responsible for the preparation and distribution of various derived products, and modelers, empirical and physical alike. Linkages are also necessary between the specialized research areas even within each element of GOALS. Achieving the linkages and coordination needed between the elements and sub–elements of GOALS may not be an easy task. The panel recommends the joint participation of scientists in GOALS projects and planning activities (including working groups and committees) to address this issue.

For example, the Atmospheric Model Intercomparison Project carries quantitative diagnostic analysis of models and observations in parallel. Similar activities should be encouraged for monsoon simulations and other major regional components of the climate system.

Scientists using diagnostic and empirical techniques should be involved in the planning of process studies and numerical experiments. The results of empirical and diagnostic studies should be used to help evaluate the accuracy and representativeness of model forecast fields, and analysis products. Furthermore, as recommended earlier, a mechanism for an ongoing dialogue between physical scientists (and institutions) who produce climate predictions and the community of users and social scientists should be established. Such activities should be coordinated by a GOALS Project Office when established as proposed in NRC (1995).

CONSORTIA AND PRINCIPAL INVESTIGATOR GROUPS

As one means of strengthening links among the six elements of GOALS, and in conjunction with an intended focus on particular geographical regions, the panel favors the development of consortia or groups of principal investigators loosely or tightly coordinated but unified by some goal or project. By combining the efforts of investigators with various specialties, consortia provide the ability to consider disparate but interlinked processes, to gather and analyze data that relates to feedbacks between processes, and to move forward with modeling and prediction of coupled phenomena.

For example, one consortium might focus on the suite of phenomena associated with short-term climate variations in the vicinity of Asia, Australia, the Indian Ocean, and the far western Pacific, and their impacts on seasonal-to-interannual climate over North America. A focus for the consortium might initially be on understanding coupled ocean–atmosphere variability in that region. Activities include the organization of a series of coordinated studies of monsoon heat sources and sinks and the underlying SST gradients, flows through straits separating the two ocean basins and their effects on the reflection of waves at the western boundaries, the role of boundary currents in the western Pacific and Indian Oceans, and the heat balance between respective basins. Improved understanding of air–sea coupling in these basins would then be built into models of global climate, with the intent of understanding the impact of the Asian–Australian monsoon system on remote locations such as North America. Another consortium might focus on the suite of problems associated with atmosphere–ocean interactions on the cold side of the tropical ocean basins. Investigators in this consortium could conduct process studies on the seasonal and temporal evolution of the SST field in these regions, with the intent of correcting existing deficiencies in understanding that lead to poor predictions of the state of the upper ocean and, in turn, of improving prediction over North America. Still another consortium might begin to consider the role of extratropical regions in seasonal-to-interannual climate. It could focus initially on the impact of meridional SST gradients on the atmosphere, conducting process studies on the local heat balance in contrasting regions, and examining the sensitivity of models to such gradients.

Activities organized and coordinated by consortia may involve any combination of empirical studies, process studies, and model simulations. In some cases, long-term observations or field experiments would be the emphasis. In other cases, modeling and consideration of applications might be the central theme. GOALS considers consortia of a number of individual investigators a powerful mechanism for achieving critical mass when addressing scientific objectives. At the same time, the efforts of individual principal investigators are a crucial component, and a significant part of the resources need to be reserved to facilitate individual research efforts.

INTERACTION WITH OTHER PROGRAMS

In recent years it has become apparent that ENSO and the monsoon systems show pronounced variability in the decadal and longer time scales. In order to investigate the modulation of seasonal-to-interannual events occurring in the climate system by features of decadal and longer duration, the panel recommends the establishment of strong links to the DecCen and ACC program components of CLIVAR. These two programs deal with natural variability on long time scales involving the "slow physics" of the system, as well as changes that could occur from increasing concentrations of atmospheric greenhouse gases and aerosols, among others.

The panel also underscores that strong interfaces need to be sustained with GEWEX because of the common interests in global energy and water cycles and their role in seasonal-to-interannual variability. This common interest of GOALS and GEWEX includes overlapping objectives aimed at improving the understanding and modeling of processes and in interpreting predictions in terms of terrestrial water resources.

Collectively, the programs discussed in this section offer a cohesive framework for studying the predictability of climate on seasonal-to-interannual time scales as envisioned by USGCRP. The functional relationship between GOALS and other important components of USGCRP are shown in Figure 12-1.

Specifically, the panel recommends partnerships between GOALS and the following programs:

1. GEWEX-GCIP, PACS, and its international program VAMOS—The former process study focuses on energy and water exchange between the atmo-

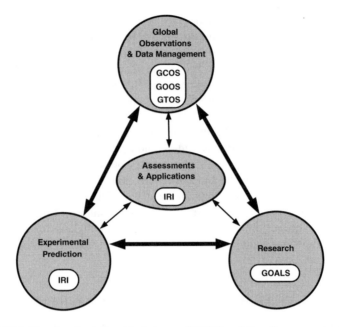

FIGURE 12-1 Functional relationship between GOALS and the other components of the U.S. Global Climate Research Program. The GOALS program will serve as the principal focus for basic research on seasonal-to-interannual time scales. GCOS, GOOS, and GTOS will provide observations on a global basis. IRI and NCEP, as well as other national and regional centers, will provide experimental prediction and assessments of seasonal-to-interannual climate variations.

sphere and the land. PACS, on the other hand, seeks linkages between variations in the tropical oceans adjacent to the Americas and seasonal-to-interannual climate variability in the Americas. Internationally, the VAMOS subprogram under CLIVAR also focuses on the variability of the American monsoons.

2. ACSYS—ACSYS oversees pilot studies, monitoring, and modeling of ocean, sea ice, and hydrological processes in the Arctic. It is important to determine the role these processes play in regional and global seasonal-to-interannual climate variability.

3. PAGES (Past Global Changes) and ARTS (Annual Records of Tropical Systems)—A joint CLIVAR-PAGES project should reconstruct the past record throughout the tropics, especially to extend the record of ENSO and monsoon variability, with sub-annual and possibly higher frequency resolution using samples from corals, tree rings, pollen, and tropical glaciers. PAGES is an International Geosphere–Biosphere Programme (IGBP) project.

4. CLIVAR-DecCen—This program, which deals with climate variability on decade-to-century time scales, should include a special emphasis from a GOALS perspective on the decadal variability of ENSO and the monsoons.

5. CLIVAR/ACC—This program deals with anthropogenic climate change and is of interest insofar as such climate change might alter shorter-term climate variability.

6. GCOS, GOOS, and GTOS—These programs are important for the organization of long-term observations involving space-based, surface, and in situ platforms and networks. Included should be coordination with the planning of IGOS.

7. WOCE—Collaboration is required, especially with reference to the upper-ocean and ocean–atmosphere interaction.

8. DOE/ARM (Atmospheric Radiation Measurement program)—ARM has three research sites located in the tropical western Pacific Ocean, in Oklahoma, and on the North Slope of Alaska. At each site, detailed radiation, cloud, and turbulent heat flux measurements are made that are invaluable for model development. Of particular interest to GOALS is a tropical western Pacific Ocean site where measurements pertain directly to processes in the warm pool.

9. NOAA-OGP projects, ACCP, IRI, SCPP, IRI, COS, and CCDD—ACCP (Atlantic Climate Change Program) is especially interested in interdecadal changes in the circulation of the Atlantic. Such changes are particularly relevant to GOALS because of the manner in which they can modulate interannual variability and, thus, predictability. The SCPP and the IRI provide predictions on seasonal-to-interannual time scales. The Climate Observing System and Climate Change and Data Detection (CCDD) concentrate on climate observing systems.

10. Other Programs—Other linkages may develop with IHDP under the International Social Science Council. They should be actively pursued within the applications and human dimensions component of GOALS.

In the United States, the panel recommends that strong linkages should be maintained between the GOALS Panel and other NRC panels and committees dealing with climate research. For applications and human dimensions, coordination is recommended with the NRC Committee on the Human Dimensions of Global Change, and its panels and working groups.

OTHER RESEARCH COORDINATION AND IMPLEMENTATION ISSUES

In addition to the partnerships proposed in the previous section, the GOALS Panel feels that there is a need to identify or establish functional mechanisms to interface GOALS with other research programs. This could include a combination of scientific workshops, meetings, and importantly joint research, model development, and data projects. The following are recommended as examples that will facilitate the coordination necessary between GOALS and other programs:

1. coordinated organization of scientific meetings (or workshops) and program development meetings and announcements of opportunity, with a focus on topics of common interest between programs, for example, a scientific meeting sponsored jointly by GOALS and GEWEX (Joint NRC panel meetings have proven useful in the past and should be encouraged.);

2. exchange of data between programs and coordinated development of data sets of relevance and value to more than one program (For example, time series of surface forcing variables for atmospheric models should be made universally available for use in sensitivity and predictability tests.);

3. exchange, testing, or coordinated development of process sub-models that are relevant to research undertaken in more than one program (For example, land–atmosphere interaction sub-models might be involved in the case of GOALS and GEWEX, and atmospheric radiation transfer sub-models in the case of GOALS, ARM, and GEWEX.);

4. coordination of GOALS with other components of CLIVAR to optimize the international effort on seasonal-to-interannual research (For example, process studies and long-term observations need to be coordinated.);

5. coordination of the timing of complementary regional studies (For example there might be simultaneous intensive observations such as between PACS and GCIP, PACS and the Land-Biosphere-Atmosphere program, and the GOALS JASMINE and other process studies, and GAME.);

6. coordination with DecCen in the development of observational efforts to understand tropical–extratropical linkages, especially in the oceans, which may be relevant to the decadal modulation of ENSO;

7. joint specification of observational systems to provide data series of value to both GOALS and DecCen (For example, an appropriate balance could be established between continuity and coverage for in situ ocean observations.);

8. coordination of the calibration and evaluation of remotely sensed variables of relevance and value to more than one program (Examples include the calibration and evaluation of TRMM and EOS data by observations from in situ monitoring under GOALS and GEWEX.);

9. interprogram support in the form of advice and guidance in the case of activities that are properly fostered within one program but also address the objectives of another program (For example, GOALS's scientists could participate in workshops that address the development and testing of coupled land–atmosphere sub-models under GEWEX or the development of atmospheric radiation transfer sub-models under ARM.);

10. joint development of coupled ocean–atmosphere–land models (For example, GOALS and GEWEX might undertake the joint development of global-scale, coupled ocean–atmosphere models that include nested mesoscale models in specific regions.);

11. joint activities addressing regional issues and the local application/interpretation of the seasonal-to-interannual predictions fostered under GOALS (For

example, GOALS and GCIP could work jointly to create a capability to predict and interpret seasonal-to-interannual variability of meteorological fields covering North America.);

12. coordinated organization of scientific meetings between GOALS and human dimensions groups to improve the use of seasonal-to-interannual predictions and to reduce human vulnerability to climate variability (Included should be the identification and provision of climate variables and products of most value to decision makers.); and

13. joint development (GOALS, IHDP) and provision of guidance in the incorporation of probabilistic climate forecasts into operational decision making processes.

GOALS PROJECT OFFICE

To coordinate and carry out the various activities mentioned in this section, the panel endorses fully the recommendations contained in the GOALS Science Plan (NRC, 1994a), namely: the establishment of a tripartite structure, with a project office (GOALS Project Office), a scientific oversight body (NRC, with its CRC, and the GOALS Panel), and a group of participating federal agencies. The federal agencies would be responsible for implementing GOALS through coordinated funding of research grants. An Interagency GOALS Project Office would serve as a focal point for the implementation of the national research effort, and the GOALS Panel (with oversight from the NRC's CRC) would provide scientific guidance for the program. The Plan also anticipated that principal investigators and "consortia" would carry out much of the actual implementation of the GOALS scientific plans. As the needs of the program dictate, the GOALS Project Office would invite groups to prepare coordinated sets of research proposals designed to address specific objectives of GOALS. The above are reiterated in the proposals on a GOALS infrastructure for U.S. participation in GOALS contained in NRC (1995). Close coordination between GOALS and the international CLIVAR is recommended by the panel through a formal link between the project offices for the two programs and an informal liaison between the GOALS Panel and the International CLIVAR Scientific Steering Group.

13

Summary of Key Recommendations and Conclusions

As a fundamental strategy, the GOALS Panel endorses the proposal contained in the Science Plan for GOALS (NRC, 1994a) to extend the domain of interest beyond the tropical Pacific Ocean (the region of primary attention during TOGA) to include the other tropical oceans and neighboring continental land masses, as well as the influence of these tropical regions on higher latitudes. However, investigations of the Pacific region need to be continued and strengthened. The panel also embraces specifically the inclusion of studies of interactions of land surface hydrological processes with the atmosphere and the oceans. To provide initial emphases for GOALS, three major regional foci are proposed:

1. the tropical Pacific Ocean;
2. the Americas and surrounding oceans; and
3. the Indian Ocean and surrounding land masses.

Due attention must be given to the global aspects of seasonal-to-interannual climate variability because of the connectivity through the atmosphere and the ocean.

For the further development of implementation plans for GOALS, the panel proposes that the program be structured around six major elements:

1. long-term observations and analyses;
2. process studies;
3. empirical and diagnostic studies;

4. modeling;
5. applications and human dimensions; and
6. data management.

A strong emphasis is placed on the maintenance and enhancement of existing observing systems, the integration of "single" and "combined" process studies, the improvement of component models of the atmosphere, the oceans, and land surface processes as well as their coupling, and a cohesive data and information management structure.

In order to involve a broader scientific community in the further planning of GOALS, the panel recommends that two working groups be formed, one on observations (GOWG) and one on modeling (GMWG). GOWG shall develop the scientific basis for a GOALS observation strategy especially with regards to spatial and temporal resolution. A key mission of the working group will be to define a strategy for space-borne observations for the duration of the GOALS program. Furthermore, GOWG will act as an interface with GEWEX and DecCen observation groups. GMWG will work towards involving a broader cross-section of the U.S. scientific community in GOALS modeling activities and encourage and coordinate a program for model intercomparison, climate system simulation, and experimental prediction. It will also act as the interface with international groups such as CLIVAR/NEG.

The panel also believes that strong interfaces are necessary between the physical science(s) communities and those of the applications, biological, and human dimensions sectors. An early and ongoing dialogue between the respective communities is recommended.

Because of the overlapping time scales of interest between the GOALS program and those of GEWEX and DecCen, strong and well coordinated interfaces are recommended between the scientists involved in the activities and projects of these programs. In order to achieve the necessary coordination between programs and between the scientific communities likely to be involved, the establishment of a GOALS Project Office is highly recommended, as well as a federal interagency coordinating mechanism. Various other detailed proposals and recommendations are also contained in the text of this report under the respective subject sections.

The panel underscores that this strategic report is not intended to be definitive or conclusive as regards the scope of GOALS. Rather, the panel strongly endorses the philosophy that the program should remain flexible to incorporate new ideas and concepts as scientific knowledge, observational and modeling technology, and so forth, improve over the duration of the 15-year program.

APPENDIX
A

Acronyms

ACC	Anthropogenic Climate Change
ACCP	Atlantic Climate Change Program
ACSYS	Arctic Climate System Study
ARGOS	ARGOS Data Collection and Location System (France)
ARM	Atmospheric Radiation Measurement program
ARTS	Annual Record of Tropical Systems
BASC	Board on Atmospheric Sciences and Climate
CCDD	Climate Change and Data Detection
CDEP	Climate Dynamics and Experimental Prediction
CENR	Committee on Environment and Natural Resources (OSTP)
CLIPS	Climate Information and Prediction System (WMO)
CLIVAR	Climate Variability and Predictability programme
CNES	Centre National d'Etudes Spatiales
COARE	Coupled Ocean–Atmosphere Response Experiment
CRC	Climate Research Committee
DecCen	Climate Variability on Decade-to-Century Time Scales
EMEX	Equatorial Mesoscale Experiment
ENSO	El Niño/Southern Oscillation
EOS	Earth Observing System
FGGE	First GARP Global Experiment (also known as the Global Weather Experiment)
GAME	GEWEX Asian Monsoon Experiment
GARP	Global Atmosphere Research Program
GCIP	GEWEX Continental-scale International Project

GCM	general circulation model
GCOS	Global Climate Observing System
GEWEX	Global Energy and Water Cycle Experiment (WCRP)
GMWG	GOALS Modeling Working Group
GOALS	Global Ocean–Atmosphere–Land System
GOOS	Global Ocean Observing System
GOS	Global Observing System
GOWG	GOALS Observation Working Group
GPCP	Global Precipitation Climatology Project
GTOS	Global Terrestrial Observing System
GTS	Global Telecommunication System (WMO)
GWE	Global Weather Experiment
IGBP	International Geosphere–Biosphere Programme (Global Change)
IGOS	Integrated Global Observing Strategy
IHDP	International Human Dimensions of Global Change Programme
IOC	International Oceanographic Commission
IPCC	Intergovernmental Panel on Climate Change
IRI	International Research Institute for Seasonal-to-Interannual Climate Prediction
ISCCP	International Satellite Cloud Climatology Project
JASMINE	Joint Air–Sea Monsoon Interaction Experiment
LBA	Land Biosphere–Atmosphere Program
LSPM	Land Surface Process Model
MSU	Microwave Sounding Unit
NASA	National Aeronautics and Space Administration
NASDA	National Space Development Agency of Japan
NCEP	National Centers for Environmental Prediction
NEG	Numerical Experimentation Group
NOAA	National Oceanic and Atmospheric Administration
NPOESS	National Polar-Orbiting Environmental Satellite System
NRC	National Research Council
NWS	National Weather Service
OGP	Office of Global Programs
OSTP/CENR	Office of Science and Technology Policy/Committee on Environment and Natural Resources
PACS	Pan American Climate Studies
PAGES	Past Global Changes
PIRATA	Pilot Research Moored Array in the Tropical Atlantic
SCPP	Seasonal-to-Interannual Climate Prediction Program
SCSMEX	South China Sea Meteorological Experiment
SST	sea surface temperature

TAO	Tropical Atmosphere Ocean (array of moorings)
TOGA	Tropical Oceans and Global Atmosphere
TRITON	Triangle Trans-Ocean buoy network
TRMM	Tropical Rainfall Measuring Mission (NASA)
USGCRP	U.S. Global Change Research Program
VAMOS	Variability of the American Monsoon Systems
VOS	Volunteer Observing Ship(s) (WMO)
WCRP	World Climate Research Programme (WMO/ICSU)
WMO	World Meteorological Organization
WOCE	World Ocean Circulation Experiment
WWW	World Weather Watch

APPENDIX
B

References and Selected Bibliography

REFERENCES

IPCC (Intergovernmental Panel on Climate Change). 1990. Climate Change: The IPCC Scientific Assessment. J.T. Houghton, G.J. Jenkins, and J.J. Ephraums, eds., Cambridge University Press, Cambridge, U. K., 365 pp.

IPCC (Intergovernmental Panel on Climate Change). 1996a. Climate Change 1995: The Science of Climate Change. J.T. Houghton, L.G. Meira Filho, B.A. Callander, N. Harris, A. Kattenberg, and K. Maskell, eds., Cambridge University Press, Cambridge, U. K., 572 pp.

IPCC (Intergovernmental Panel on Climate Change). 1996b. Economic and Social Dimensions of Climate Change. J.P. Bruce, H. Lee, and E.F. Haites, eds., Cambridge University Press, Cambridge, U. K., 448 pp.

NRC (National Research Council). 1994a. (GOALS) Global Ocean–Atmosphere–Land System for Predicting Seasonal-to-interannual Climate. National Academy Press, Washington, D.C., 103 pp.

NRC (National Research Council). 1994b. Ocean–Atmosphere Observations Supporting Short-term Climate Predictions. National Academy Press, Washington, D.C., 51 pp.

NRC (National Research Council). 1995. Organizing U.S. Participation in GOALS (Global Ocean–Atmosphere–Land System). GOALS Panel. National Academy Press, Washington, D. C., 8 pp.

NRC (National Research Council). 1996. Learning to Predict Climate Variations Associated with El Niño and the Southern Oscillation: accomplishments and legacies of the TOGA program. National Academy Press, Washington, D. C., 171 pp.

NRC (National Research Council). 1997. BITS of POWER: Issues in global access to scientific data. National Academy Press, Washington, D.C., 235 pp.

WCRP (World Climate Research Programme). 1995. Climate Variability and Predictability (CLIVAR) Science Plan. World Meteorological Organization, Geneva, Switzerland, WCRP-89, WMO/TD-No. 690.

WMO (World Meteorological Organization). 1995a. Plan for the Global Climate Observing System (GCOS), Version 1.0, May 1995. World Meteorological Organization, Geneva, Switzerland, GCOS-14, WMO/TD-No. 681, 49 pp.

WMO (World Meteorological Organization). 1995b. GCOS/GTOS Plan for Terrestrial Climate-related Observations, Version 1.0, November 1995. World Meteorological Organization, Geneva, Switzerland, GCOS-21, WMO/TD-No. 721, UNEP/EAP.TR/95-07

SELECTED BIBLIOGRAPHY

Anderson, D.L.T., and J.P. McCreary. 1985. Slowly propagating disturbances in a coupled ocean–atmosphere model. J. Atmos. Sci. 42:615-629.

Atlas, R., N. Wolfson, and J. Terry. 1993. The effect of SST and soil moisture anomalies on GLA model simulations of the 1988 U. S. summer drought. J. Climate 6:2034-2048.

Barnett, T.P., N. Graham, M. Cane, S. Zebiak, S, Dolan, J. O'Brien, and D. Leger. 1988. On the prediction of the El Niño of 1986-1987. Science 241:192-196.

Battisti, D.S., A.C. Hirst, and E.S. Sarachik. 1989. Instability and predictability in coupled atmosphere–ocean models. Phil. Trans. R. Soc. London A329:237-247.

Branstator, G.W. 1995. Organization of storm track anomalies by recurring low-frequency circulation anomalies. J. Atmos. Sci. 52:207-226.

Busalacchi, A.J., K. Takeuchi, and J.J. O'Brien. 1983. Interannual variability of the equatorial Pacific—revisited. J. Geophys. Res. 88:7551-7562.

Cane, M.A. 1991. Forecasting El Niño with a geophysical model. Pp. 345-369 in Teleconnections Linking Worldwide Climate Anomalies, M.H. Glantz, R.W. Katz, and N. Nicholls, eds., New York: Cambridge University Press.

Chapman, W.L. and J.E. Wals. 1993. Recent variations of sea ice and air temperature in high latitudes. Bull. Am. Meteorol. Soc. 74:33-47.

Chen, D., S.E. Zebiak, A.J. Busalacchi, and M.A. Cane. 1995. An improved procedure for El Niño forecasting: Implications for predictability. Science 269:1699-1702.

Cole, J.E., R.G. Fairbanks, and G.T. Shen. 1993. Recent variability in the Southern Oscillation: Isotopic results from a Tarawa atoll coral. Science 260:1790-1793.

Delworth, T. and S. Manabe. 1989. The influence of soil wetness on near-surface atmospheric variability. J. Climate 2:1447-1462.

Dunbar, R.B., G.M. Wellington, M.W. Colgan, and P.W. Glynn. 1994. Eastern Pacific sea surface temperature since 1600 AD: The $\delta^{18}O$ record of climate variability in Galapagos corals. Paleoceanography 9:291-315.

Fenessy, M. and J. Shukla. 1992. Influence of initial soil wetness on prediction of global atmospheric circulation and rainfall. Paper presented at Second International Conference on Modeling of Global Climate Change and Variability, Hamburg, Germany, September 7-11, 1991.

Folland, C.K., T.N. Palmer, and D.E. Parker. 1986. Sahel rainfall and worldwide sea temperatures, 1901-85. Nature 320:602-607.

Frederiksen, J.S. and P.J. Webster. 1988. Alternative Theories of Teleconnections and Low Frequency Fluctuations. Review of Geophysics 26:459-494.

Gates, W.L. 1992. AMIP: The Atmospheric Model Intercomparison Project. Bull. Am. Meteor. Soc. 73:1962-1970.

Geisler, J.E., M.L. Blackmon, G.T. Bates, and S. Mu. 1985. Sensitivity of January climate response to the magnitude and position of equatorial Pacific sea surface temperature anomalies. J. Atmos. Sci. 42:1037-1049.

Glantz, M.H., R.W. Katz, and N. Nicholls, eds., 1991. Teleconnections linking worldwide climate anomalies. Cambridge University Press, 535 pp.

Graham, N.E., J. Michaelsen, and T.P. Barnett. 1987a. An investigation of the El Niño/Southern Oscillation cycle with statistical models. Part 1: Predicted field characteristics. J. Geophys. Res. 92:14251-14270.

Graham, N.E., J. Michaelsen, and T.P. Barnett. 1987b. An investigation of the El Niño/Southern Oscillation cycle with statistical models. Part 2: Model results. J. Geophys. Res. 92:14271-14289.

Harrison, D.E. and N.K. Larkin. 1997. Darwin sea level pressure, 1876-1996: Evidence for climate change? Geophys. Res. Lett. 24:1779-1782.

Hasselman, K. 1976. Stochastic Climate Models. Part 1. Theory. Tellus 28:473-485.

Hastenrath, S. 1990. Prediction of Northeast Brazil Rainfall Anomalies. J. Climate 3:893-904.

Hayes, S.P., L.J. Mangum, J. Picaut, A. Sumi, and K. Takeuchi. 1991. TOGA-TAO: A moored array for real-time measurements in the tropical Pacific Ocean. Bull. Amer. Meteor. Soc. 72:339-347.

Ji, M., A. Kumar, and A. Leetmaa. 1994. A multiseason climate forecast system at the National Meteorological Center. Bull. Am. Meteor. Soc. 75:569-577.

Ji, M., A. Leetmaa, and V.E. Kousky. 1995. An ocean analysis system for seasonal to interannual climate studies. Mon. Wea. Rev. 123:460-481.

Kalnay, E., M. Kanamitsu, R. Kistler, W. Collins, D. Deaven, L. Gandin, M. Iredell, S. Saha, G. White, J. Woollen, Y. Zhu, M. Chelliah, W. Ebisuzaki, W. Higgins, J. Janowiak, K.C. Mo, C. Ropelewski, A. Leetmaa, R. Reynolds, and R. Jenne. 1996. The NCEP/NCAR Reanalysis Project. Bull. Am. Meteor. Soc. 77:437-471.

Karl, T. R., ed. 1996. Long-term climate monitoring by the Global Climate Observing System. International meeting of experts, Asheville, North Carolina, U. S. A.. Reprinted from Climate Change Vol 31, Nos. 2-4, 1995. Kluwer Academic Publisher, The Netherlands, 648 pp.

Karl, T.R., P.Y. Groisman, R.R. Heim, Jr., and R.W. Knight. 1993. Recent variations of snowcover and snowfall in North America and their relation to precipitation and temperature variability. J. Clim. 6:1327-1344.

Karl, T.R., G. Kukla, and J. Gavin. 1984. Decreasing diurnal temperature range in the United States and Canada from 1941-1980. J. Clim. Appl. Met. 23, 23:1489-1504.

Knutson, T.R. and S. Manabe. 1995. Time-mean response over the tropical Pacific to increased CO_2 in a coupled ocean–atmosphere model. J. Clim. 8:2181-2199.

Knutson, T.R., S. Manabe, and D. Gu. 1997. Simulated ENSO in a global coupled ocean–atmosphere model: multidecadal amplitude modulation and CO_2 sensitivity. J. Clim. 10:138-161.

Latif, M., A. Sterl, E. Maier-Reimer, and M.M. Junge. 1993. Structure and predictability of the El Niño Oscillation phenomenon in a coupled ocean–atmosphere general circulation model. J. Clim. 6:700-708.

Lau, K.M., C.H. Sui, and T. Nakazawa. 1989. Dynamics of westerly wind burst, supercloud clusters, 30-60 day oscillations and ENSO: A unifying view. J. Meteorol. Soc. Japan 67:205-219.

Lukas, R., 1988: Interannual fluctuations of the Mindanao current inferred from sea level. J. Geophys. Res. 93: 6744-6748.

Lukas, R., P.J. Webster, M. Ji, and A. Leetmaa. 1995. The large scale context for the TOGA COARE coupled ocean–atmosphere response experiment. Meteorol. and Atmos. Physics 56:3-16.

McPhaden, M.J. 1995. The Tropical Atmosphere–Ocean array is completed. Bull. Am. Meteorol. Soc. 76:739-741.

Meehl, G.A. and W.M. Washington. 1996. El Niño like climate change in a model with increased atmospheric CO_2 concentrations. Nature 382:56-60.

Meehl, G.A., G.W. Branstator, and W.M. Washington. 1993. Tropical Pacific interannual variability and CO_2 climate change. J. Clim. 6:42-63.

Molteni, F., L. Ferranti, T.N. Palmer and P. Viterbo. 1993. A dynamical interpretation of the global response to equatorial Pacific SST anomalies. J. Clim. 6:777-795.

Moura, A.D. 1992. International Research Institute for Climate Prediction: A proposal. NOAA Office of Global Programs, Silver Spring, Maryland, 51 pp.

Nicholls, N., B. Lavery, C. Frederiksen, and W. Drosdowsky. 1996. Recent apparent changes in relationships between the El Niño/Southern Oscillation and Australian rainfall and temperature. Geophys. Res. Lett. 23:3357-3360.

Palmer, T.N. 1993. A nonlinear dynamical perspective on climate change. Weather 48:314-326.

Philander, S.G.H, R.C. Pacanowski, N.K. Lau, and M.J. Nath. 1992. Simulation of the ENSO with a global atmospheric GCM coupled to a high-resolution, tropical Pacific Ocean GCM. J. Clim. 5:308-329.

Patz, J.A., P.R. Epstein, T.A. Burke and J.M. Balbus. 1996. Global climate change and emerging infectious diseases. JAMA 275:217-223.

Rajagopalan, B., U. Lall, and M.A. Cane, 1997. Anomalous ENSO occurrences: An alternative view. J. Clim. 10:2351-2357.

Rasmusson, E.M. and K.C. Mo. 1993. Linkages between 200-mb tropical and extratropical anomalies during the 1986-1989 ENSO cycle. J. Clim. 6:595-616.

Rassmusson, E.M., X. Wang, and C.E. Ropelewski. 1990. The biennial component of ENSO. J. Mar. Sys. 1:71-96.

Ropelewski, C.F. and M.S. Halpert. 1989. Global and regional scale precipitation patterns associated with the El Niño/Southern Oscillation. Mon. Wea. Rev. 115:1606-1626.

Sarachik, E.S. 1990. Predictability of ENSO. Pp. 161-171 in Ocean Climate Interaction, M. E. Schelisinger, ed., Norwell, Massachusetts: Kluwer Academic Publisher.

Shukla, J. 1987. Long range forecasting of monsoons. Pp. 523-547 in Monsoons, J.S. Fein and P. L. Stephens, eds., New York: J. Wiley & Sons, Ltd.

Tomas, R. and P.J. Webster. 1997. On the location of the intertropical convergence zone and near-equatorial convection: The role of inertial instability. Quart. J. Roy. Met. Soc. 123:1445-1482.

Trenberth, K.E. 1995. Atmospheric circulation climate changes. Climatic Change 31:427-453.

Trenberth, K.E. 1996. El Niño-Southern Oscillation, Chapter 6 of Climate Change: Developing Southern Hemisphere Perspectives. T. Giambelluca and A. Henderson-Sellers, eds., John Wiley & Sons, Ltd., 145-173.

Trenberth, K.E. 1997. Short-term climate variations: Recent accomplishments and issues for future progress. Bull. Amer. Meteor. Soc. 78:1081-1096.

Trenberth, K.E. and T.J. Hoar. 1996. The 1990-1995 El Niño-Southern Oscillation event: Longest on record. Geophys. Res. Lett. 23:57-60.

Trenberth, K.E., G.W. Branstator, D. Karoly, A. Kumar, N.C. Lau, and C. Ropelewski. 1996. Global atmospheric diagnostics and modeling for TOGA. (special TOGA issue)

Unninayar, S. and K.H. Bergman. 1993. Earth system modeling in the mission to planet Earth era. NASA/MTPE monograph, 133 pp.

Wang, B. 1995. Interdecadal changes in El Niño onset in the last four decades. J. Clim. 8:267-285.

Webster, P.J. 1982. Seasonality of atmospheric response to sea-surface temperature anomalies. J. Atmos. Sci. 39:29-40.

Webster, P.J. and R.A. Houze, Jr. 1991. The Equatorial Mesoscale Experiment, EMEX. Bull. Am. Met. Soc. 72:1481-1505.

Webster, P.J. and R. Lucas. 1992. TOGA-COARE: The coupled ocean–atmosphere response experiment. Bull. Am. Met. Soc. 73:1377-1416.

Webster, P. 1994. The role of hydrological processes in ocean–atmosphere interaction. Rev. Geophys. 32:427-476.

Webster, P.J. 1995. The annual cycle and the predictability of the tropical coupled ocean–atmosphere system. Meteorol. Atmos. Physics 56:33-55.

Webster, P.J., T. Palmer, M. Yanai, V. Magana, J. Shukla, and A. Yasunari. 1998. The Monsoon: processes, predictability and the prospects for prediction. J. Geophys. Res. special issue (in press)

Zhang, Y., J.M. Wallace, and D.S. Battisti. 1997. ENSO-like interdecadal variability: 1900-93. J. Clim. 10:1004-1020.

Zebiak, S.E. and M.A. Cane. 1987. A model El Niño/Southern Oscillation. Mon. Wea. Rev. 115:2262-2278.